ULTIMATE
HARLEY-DAVIDSON

ULTIMATE
HARLEY-DAVIDSON

Hugo Wilson

LONDON, NEW YORK, MUNICH,
MELBOURNE, DELHI

THIS EDITION

PROJECT EDITOR Alison Sturgeon
PROJECT ART EDITOR Tannishtha Chakraborty
US SENIOR EDITOR Shannon Beatty
US EDITOR Jenny Siklos
MANAGING EDITOR Laura Buller
MANAGING ART EDITOR Karen Self
ART DIRECTOR Phil Ormerod
ASSOCIATE PUBLISHER Liz Wheeler
REFERENCE PUBLISHER Jonathan Metcalf
PRE-PRODUCTION CONTROLLER Ben Marcus
JACKET DESIGNER Mark Cavanagh
JACKET EDITOR Manisha Majithia

PREVIOUS EDITIONS

SENIOR EDITORS Peter Jones, Neil Lockley
SENIOR ART EDITOR Heather McCarry
MANAGING EDITORS Anna Kruger, Adele Hayward
MANAGING ART EDITORS Karen Self,
Steve Knowlden
US EDITORS Christine Heilman,
Chuck Wills, Gary Werner
DEPUTY ART DIRECTOR Tina Vaughan
DTP DESIGNERS Rob Campbell, Rajen Shah
PICTURE RESEARCHER Jamie Robinson
CATEGORY PUBLISHER Stephanie Jackson
PRODUCTION CONTROLLERS Louise Daly,
Sarah Sherlock
PHOTOGRAPHY BY Dave King

PREVIOUS EDITIONS PACKAGED FOR
DORLING KINDERSLEY BY

ART EDITORS Mark Johnson Davies, Tracy Miles
EDITORS Phil Hunt

Revised American Edition, 2013

Published in the United States by
DK Publishing, Inc.
375 Hudson Street
New York, New York 10014

03 04 05 06 07 10 9 8 7 6 5 4 3 2 1

Reproduced by Colourscan, Singapore
Printed in China by South China

Discover more at
www.dk.com

PAGE 2:
1915 Board-Track Racers
PAGE 3:
1999 Twin Cam Engine

CONTENTS

JOE WALTER ON A 1915 KT BOARD RACER

MODEL S RACER

THE PANHEAD ENGINE

VRSCA V-ROD

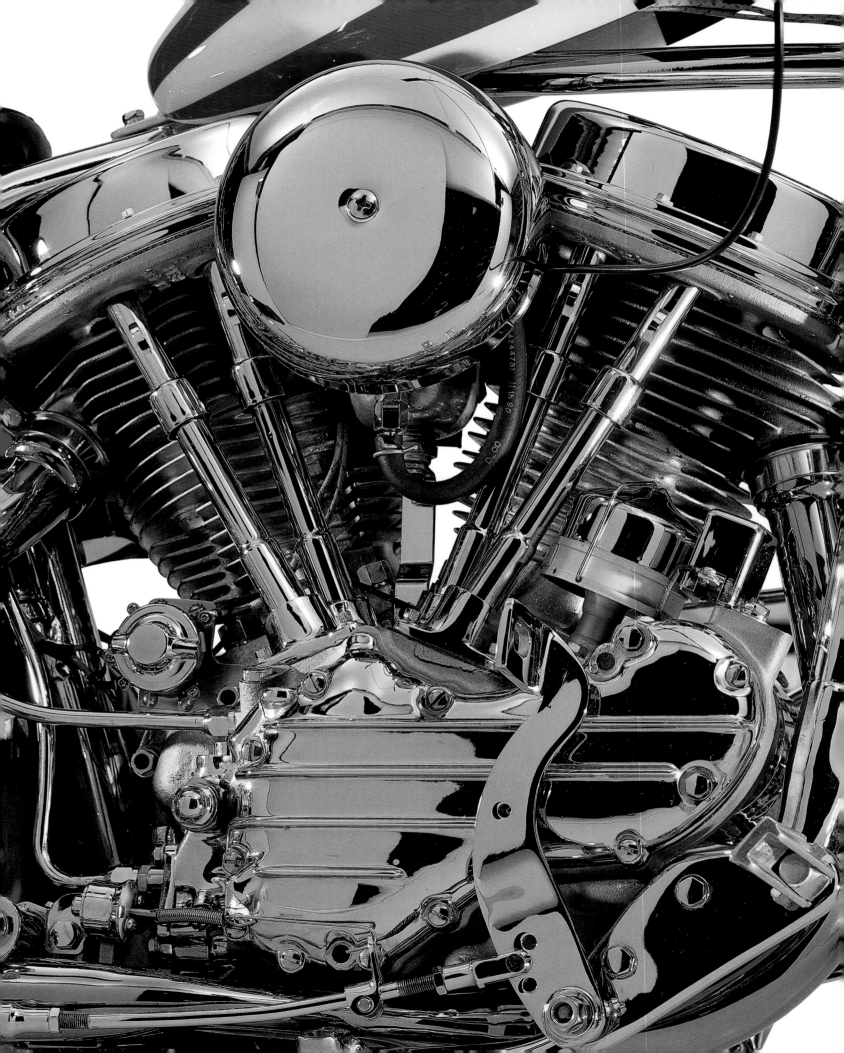

Foreword

Harley-Davidson for many people is synonymous with the motorcycle. This honor is not without merit. Harley has done a great deal for the motorcycle industry as well as motorcycle racing and every Harley-Davidson has a story. Whether it was ridden to Sturgis, raced at Daytona, or used for milk delivery, these motorcycles served their purpose well and with a flare that is Harley-Davidson. The roar of a Harley is pure motorcycle.

The Harleys of the Barber Vintage Motorsports Museum range from board-track racers to a Captain America replica. There are military Harleys, Italian Harleys, touring Harleys, a Harley-powered midget racer, a Knucklehead, and even Roger Reiman's KR that raced on the beach at Daytona and then went on to win the inaugural 200 at the speedway.

Within our research library at the museum of over 2,500 books we have more than 190 just devoted to Harley-Davidson. This one will be a welcome addition and shining star. We are proud to have been a part of it.

George Barber Jr.

George W. Barber
The Barber Vintage Motorsports Museum,
Birmingham, Alabama.
www.BarberMuseum.org

Author's Preface

People remember their first Harley-Davidson: the first time they see one, the first time they hear one, the first time they ride one. They get to you like that. Harleys look different, sound different, and certainly ride different from other motorcycles. Their extraordinary appeal makes small boys dream and grown men save. Harley-Davidson has been making motorcycles for over 100 years. The company that started as four young men hand-building simple machines in a wooden shed is now one of the world's most famous brands. You will find some of their most memorable bikes on the pages that follow.

HUGO WILSON

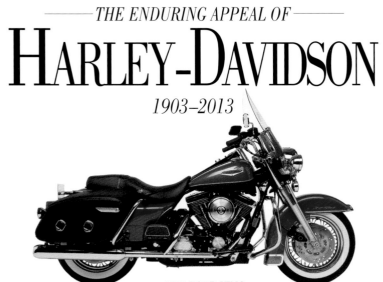

THE ENDURING APPEAL OF

HARLEY-DAVIDSON

1903–2013

1998 ROAD KING

THE STORY OF Harley-Davidson is enthralling.
It began in an era from which only the
most determined and farsighted have
survived, and in many ways Harley's story
reflects the culture in which the company has
flourished—unparalleled growth punctuated
by periods of decline, but always operating
with a strong spirit of survival.

AN AMERICAN DREAM
*Harley's interpretation of the American Dream
is one of the most coveted symbols of freedom and
independence for bikers around the world.*

THE ENDURING APPEAL OF
HARLEY-DAVIDSON

HARLEY-DAVIDSON HAS BEEN AROUND ALMOST as long as the motorcycle, and longer than the motorcycle has been a half sensible form of transportation. In 1903 Harley-Davidson was just four friends in their twenties, all motorcycle fanatics at a time when the only people who owned motorcycles were the seriously rich or those who actually made them. They could not have foreseen the legacy they would leave.

1903 TO 1919

At the turn of the 20th century, America was a place for adventure and challenge. William S. Harley and brothers Arthur, William, and Walter Davidson embarked on their own adventure of becoming motorcycle manufacturers. While some pioneer manufacturers bolted an existing engine into a bicycle frame, the four young men from Milwaukee did it the difficult way. In 1903 they constructed an engine from scratch and redesigned the frame to make it stronger and more suitable for its new role. The first bikes were put together in a 10ft x 15ft (3m x 4.5m) shed on the premises of the Davidsons' family home, though a new "factory" was built in 1906 that measured 28ft x 80ft (8.5m x 24.5m). William Harley interrupted his work at Harley-Davidson to study automotive engineering at the University of Wisconsin where, rumor has it, he developed the sprung fork as a college project. He raised financial support for his studies by waiting tables, and for all four of them the motorcycle project was only a part-time interest until they moved to the larger premises in 1906.

Conditions were rough on the roads of the US in the early 20th century, and the distances between towns were considerable. If a motorcycle was to be a viable machine rather than an amusement, it had to be reliable, tough, practical, and powerful. The fledgling Harley-Davidson company understood this from the start, and even its earliest machines were more robust than most other bikes on the market. While its competitors came and went, Harley continued to develop its machines and enhance its reputation. Production numbers leapt from three machines in 1903 to 50 in 1906, just over 1,100 in 1909, and double that the following year. By 1919 the company was producing over 23,000 bikes a year and was the second biggest manufacturer in the US behind the mighty Indian operation. Impressive figures considering that the Harley-Davidson Motor Company was only established in 1907.

THE FIRST HARLEY-DAVIDSON FACTORY IN MILWAUKEE

Threat from the car
The biggest competition to Harley-Davidson, and to the other American motorcycle manufacturers, came from Henry Ford and his Model T car. Following its introduction in 1908, the car gradually became cheaper as mass production developed, and it soon became less expensive than all but the most rudimentary of motorcycles. Those who were prepared

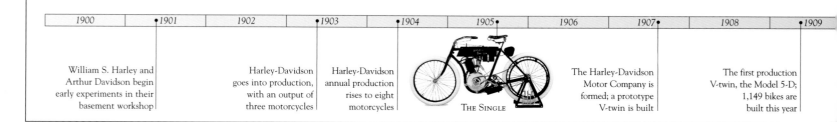

1900	1901	1902	1903	1904	1905	1906	1907	1908	1909
	William S. Harley and Arthur Davidson begin early experiments in their basement workshop	Harley-Davidson goes into production, with an output of three motorcycles	Harley-Davidson annual production rises to eight motorcycles		THE SINGLE		The Harley-Davidson Motor Company is formed; a prototype V-twin is built		The first production V-twin, the Model 5-D; 1,149 bikes are built this year

THE FOUNDING FATHERS ARTHUR DAVIDSON, WALTER DAVIDSON, WILLIAM A. DAVIDSON, AND WILLIAM S. HARLEY

to spend cash on a motorcycle were either enthusiasts or police departments, who soon recognized that high-performance motorcycles were rather useful for catching misbehaving motorists in low-performance automobiles. For motorcycle manufacturers, it was a case of either developing new technologies of their own to take the motorcycle forward, or simply going out of business. And most of Harley-Davidson's competitors did.

Fortunately, Harley made significant technological advances in its first few years. The sprung fork appeared in 1907, magneto ignition in 1909, mechanical inlet valves in 1911, and chain drive in 1912. Mechanical oil pumps appeared in 1915, along with electric lighting and a three-speed gearbox, by which time one could argue that the motorcycle had achieved a practical form and all further developments were improvements rather than breakthroughs.

Birth of the V-twin

But the most important development for Harley-Davidson occurred between 1907 and 1911. In 1907 the company built its first experimental V-twin-engined machine. Two years later a V-twin was listed as part of the model line, but disappeared the following year, suggesting that it had not been perfected. The V-twin returned in 1911, and is still in

HARLEY-DAVIDSON SIDECAR IN ACTION DURING WORLD WAR I

production nine decades later. Though Harley's bar and shield trademark first appeared in 1910, it is the 45° V-twin that remains a more potent symbol of the company. With hindsight, it's easy to sound wise. So when Harley introduced a new lightweight flat-twin in 1919, we can see that it could never have been a success. But in 1919 the company couldn't have realized that its future prospects would be so entwined with the V-twin engine.

The years leading up to the end of that decade were a period of extraordinary growth. By 1914, the tiny workspace of 1903 had grown to an area covering 2,424,479ft^2 (225,234m^2) and the number of employees had risen from four to over 1,500. Harley's bikes had quickly gained a name for being fast and dependable and the company set up a factory race team in 1915 which had immediate success. In addition, Harley was given an official seal of approval by supplying over 20,000 bikes to Allied forces during World War I.

A combination of innovative engineering and shrewd business practice meant that in less than 15-years Harley-Davidson had established itself as one of the world's leading motorcycle manufacturers.

EARLY HARLEY-DAVIDSON HILL-CLIMBER

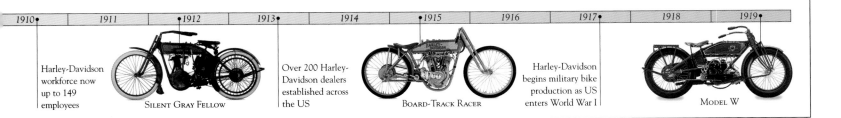

1910	1911	1912	1913	1914	1915	1916	1917	1918	1919

Harley-Davidson workforce now up to 149 employees

SILENT GRAY FELLOW

Over 200 Harley-Davidson dealers established across the US

BOARD-TRACK RACER

Harley-Davidson begins military bike production as US enters World War I

MODEL W

1920 TO 1939

The period between the wars saw production slip back in terms of volume as the domestic market shrank. Consequently Harley went on a successful search for export markets—as well as Europe and the British empire, machines were also exported to Japan. At home, dealers were given increased support, advertising budgets soared, and credit programs were developed. Everything possible was done to persuade the fence-sitter that they could and should buy a new Harley-Davidson. Though the factory in Milwaukee wasn't running at anything like the capacity that it had been, it was still doing far better than any other manufacturer. In the mid-1920s Harley-Davidson overtook Indian as the biggest motorcycle manufacturer in the United States, and for a time it was the biggest in the world.

In motorcycle racing, the board tracks that had provided the main spectacle in the early years of the century—and that had given Harley a number of notable victories—were falling from favor, partly as a result of some appalling accidents. By the mid-1920s, dirt-track racing had become the next big thing and was growing in popularity.

For the bike-buying public, the range of Harleys on offer was changing all the time. The W-series flat-twins, introduced in 1919, were dropped in 1923, having failed on the American market. Small-capacity, single-cylinder machines such as the A and B were brought in to fill a similar market niche. Sales were steady until the models were dropped in the early 1930s.

Far more significant were the new 45cu. in. side-valve V-twins, which were launched in 1929. These were built to compete with Indian and Excelsior in what was an expanding area of the market. In 1932, a three-wheeled version of the side-valve V-twin was built— the Servi-Car became popular with the police and production

continued until the 1970s. When new 74cu. in. side-valve V-twins were introduced in 1930, it signaled the end for Harley's long association with the inlet-over-exhaust engine which had been used on Harleys since its first machines.

The depression bites

The Wall Street Crash of 1929 changed the American economy overnight, and the depression that followed had a severe impact on Harley-Davidson and its competitors. The Excelsior-Henderson company stopped motorcycle production in 1931, and in 1933 Harley production slumped to less than 4,000 machines, its lowest figure since 1910. The company survived the Great Depression intact partly because of the family ownership of the firm; quite

F.A. LONGMAN ON A WINNING HARLEY-DAVIDSON

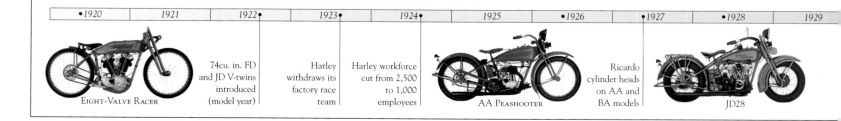

•1920	1921	1922•	1923•	1924•	1925	•1926	•1927	•1928	1929
EIGHT-VALVE RACER		74cu. in. FD and JD V-twins introduced (model year)	Harley withdraws its factory race team	Harley workforce cut from 2,500 to 1,000 employees	AA PEASHOOTER		Ricardo cylinder heads on AA and BA models	JD28	

THE VICTORIOUS 1920 HARLEY-DAVIDSON FACTORY RACE TEAM

Though the general economic climate had improved by the mid-1930s, Harley's situation was still uncertain, so the decision to release what is arguably the most significant model in Harley's history was a risky one. The technologically advanced "Knucklehead" had a recirculating lubrication system, a four-speed gearbox, and overhead-valves with hemispherical combustion chambers. This bike is the direct granddaddy of today's big-twins, but its influence was more than just mechanical. The 1930s was the decade of streamlining, when aerodynamics wasn't a science but a statement. It was about optimism in future technology.

THE SIDECAR WAS A GREAT SUCCESS IN THE 1920s, WITH BOTH LEISURE AND RACING MODELS' POPULARITY

simply, Harley-Davidson didn't have to humor its shareholders. The families who owned the company had to roll up their sleeves, get on with the job, and wait for the economic climate to improve.

The depression was giving motorcycle racing a hard time, too. In 1934 a new racing class was introduced that encouraged production-based machines and amateur racers back onto the tracks. Class C was a big hit, and it soon became the most important racing category in America.

A classic is unveiled

There was a bright side to the era. The 1930s Art Deco movement, with its fresh use of color and styling, inspired Harley-Davidson to move away from the green color schemes it had been using as its paint finishes since 1917. And one major new model introduced in 1936 embodied the movement's style and boldness—the 61E "Knucklehead."

And the 61E epitomized the streamlined look, an integrated style that set the tone for the modern motorcycle.

In the run up to World War II, Harley-Davidson was in a comfortable position. It had a solid model line, a good reputation, a wide dealer network, and a secure financial base. It had grown from strength to strength and, most importantly, had survived the worst economic collapse on record.

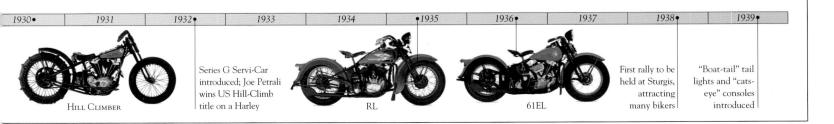

1930	1931	1932	1933	1934	1935	1936	1937	1938	1939

HILL CLIMBER

Series G Servi-Car introduced; Joe Petrali wins US Hill-Climb title on a Harley

RL

61EL

First rally to be held at Sturgis, attracting many bikers

"Boat-tail" tail lights and "cats-eye" consoles introduced

1940 TO 1959

By the time the United States entered World War II after the Japanese attack on Pearl Harbor in December 1941, the economy had already adjusted itself to the effects of the conflict. Harley-Davidson's civilian motorcycle production had been put on hold earlier in 1941 as the company geared up for the war effort, and during the following four years over 90,000 military machines were supplied to the Allies. The vast majority of these were side-valve WLAs, their widespread use on the battlefields of Europe helping to advertise the Harley-Davidson name worldwide.

Harley emerged from World War II in good shape, though production and supply of civilian machines did not reach full capacity again until 1947. The war had other consequences for Harley-Davidson and the American motorcycle scene in general. The availability of cheap ex-military motorcycles, and the fact that there were large numbers of demobbed military personnel looking for excitement, gave rise to the trend for customizing motorcycles. Standard bikes were stripped of all extraneous parts to improve handling and increase performance. These bikes became known as "Bobbers," and were the forerunners of the later choppers.

One incident that occurred just after the war damaged the reputation of motorcycling. In July 1947, a large group of bikers known as "The Booze Fighters" met up in the town of Hollister,

A WLA OFFERS PROTECTION TO ITS ARMED DESPATCH RIDER

California, and indulged in some exuberant behavior. This was subsequently reported as a full-scale riot that put the lives and property of the townsfolk in danger. Motorcyclists were now seen as hell-raisers in black leather, a label that took some time to shift. The events at Hollister became the basis of the film *The Wild One*, which starred Marlon Brando (though he actually rode a Triumph in the film).

A changing scene

Another side-effect of the war was that Harley acquired the design for a 125cc two-stroke machine from the German company DKW as part of war reparations. Production of the small bike began in 1948 and, though it was Harley's first two-stroke and a radical departure from its traditional machines, its later derivatives such as the Hummer sold well in prosperous 1950s America. The other major new model for 1948 was an updated version of the "Knucklehead" big-twin, dubbed the "Panhead" because of the appearance of its rocker covers. The following year the arrival of Harley's first hydraulically damped telescopic forks heralded the appearance of the "Hydra-Glide."

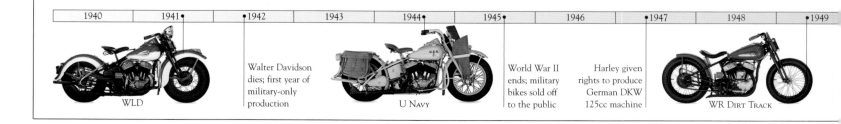

1940	1941	1942	1943	1944	1945	1946	1947	1948	1949

WLD

Walter Davidson dies; first year of military-only production

U Navy

World War II ends; military bikes sold off to the public

Harley given rights to produce German DKW 125cc machine

WR DIRT TRACK

Away from Harley's big-twin development, things were moving apace in the motorcycle market. The arrival of British motorcycles in the United States from the late 1940s saw the first real overseas threat to Harley-Davidson's dominant market position. These models were lighter, quicker, and handled better than the traditional Harleys, with an emphasis on performance and looks rather than

HARLEYS SOLD IN LARGE NUMBERS TO POLICE FORCES IN THE 1950s (ABOVE), WHILE SMALLER BIKES (RIGHT) WERE TOUTED AS THE NEW FUN WAY TO GET AROUND

rugged durability and long-distance comfort. Harley responded in 1952 with the new side-valve K Sport, but the limitations of the flathead engine layout meant that it wasn't until the bike gained overhead-valves and was christened the Sportster in 1957 that it became a real success. It is now the longest surviving motorcycle model in the world.

In a further development, the Indian company, which had been building motorcycles in America since 1901, ended production in 1953. This was the year that Harley-Davidson

was celebrating its 50th anniversary, and Indian's closure left Harley as the only significant manufacturer still operating in the United States. Indian had once been the biggest manufacturer in the world, with an enviable reputation for quality and innovation. The rival company's collapse, with Harley in rude health, underlined the strength of the foundations at the Milwaukee company.

A VERY YOUNG WILLIE G. DAVIDSON

Star treatment

Harley-Davidson bikes were getting more high-profile, and Elvis Presley was one of a number of stars who were prepared to declare themselves Harley owners (though it didn't stop him riding a Honda in the 1964 film *Roustabout*). In 1956 he even appeared on the cover of Harley's magazine, *The Enthusiast*, aboard his KH Sport twin. Others who were used to sell the Harley name included Clark Gable, Tyrone Power, and Roy Rogers. The association between the Harley brand and contemporary "stars" has continued through to the present day.

The late 1950s saw a couple of significant new models being unveiled. In 1957 the unit-construction Sportster appeared, and at last Harley had a model that could match the performance and looks of the British imports. Then in 1958 the big-twin got rear suspension and was renamed the "Duo-Glide," a model that established the blueprint for the big Harley tourer; comparing the profile of a "Duo-Glide" with a modern Harley, it is hard to tell them apart. Given that the Sportster has remained essentially the same since its introduction, it can be argued that Harley's unique styling was established in the 1950s.

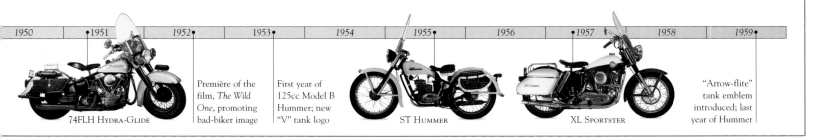

1950	1951	1952	1953	1954	1955	1956	1957	1958	1959

74FLH HYDRA-GLIDE

Première of the film, *The Wild One*, promoting bad-biker image

First year of 125cc Model B Hummer; new "V" tank logo

ST HUMMER

XL SPORTSTER

"Arrow-flite" tank emblem introduced; last year of Hummer

1960 TO 1979

In 1960, America was booming and motorcycle sales were strong, but Harley-Davidson wasn't getting as much of the action as it would have liked. Part of the reason was that the scooter market had taken off in the affluent 1950s. Period sales brochures showed scooters being piloted by well-scrubbed college kids in checked shirts. This market was fed by domestic producers and also by imported machines from Italy and Germany. It was unfortunate that when Harley-Davidson chose to enter the market in 1960 with its Topper scooter, the market had started to shrink.

Meanwhile, British-built 500 and 650cc twins continued to sell well, and the arrival of Honda (in 1959) and other Japanese manufacturers changed the market again. Their marketing campaigns, targeting people who hadn't previously thought of buying a motorbike, expanded the market for small motorcycles. In addition, they knew that some of the people who started on a small bike would soon be looking for bigger machines.

Harley was able to offer a couple of small bikes of its own for, as well as the Topper, the company had developed a 165cc lightweight two-stroke based on the 125 which first appeared in 1948. Ultimately, however, it was not enough and Harley must have figured that it didn't have the expertise or inclination to compete for small-bike sales without outside help. In 1960 it bought a 50 percent stake in the Italian company Aermacchi and instantly acquired a selection of small-capacity machines. Aermacchi's bikes were re-badged as Harley-Davidsons to immediately increase the Harley range. Unfortunately, Harley dealers were even more suspicious of the Italian-built machines than they had been of the scooter, and sales were disappointing.

HARLEY'S MERGER WITH AMF BOOSTED THE COMPANY'S FINANCIAL POSITION

PETER FONDA AND DENNIS HOPPER TAKE TO THE ROAD ON THEIR HARLEY CHOPPERS IN THE 1969 FILM *EASY RIDER*

The stakes are raised

The arrival of high-spec Japanese imports had another consequence. People wondered why, if a Japanese 125cc machine could have an electric starter, an American 1200cc machine could not. Harley's response was to fit an electric starter to its FL models from 1965 to create the Electra Glide, possibly the most famous motorcycle model ever built. It even had a film named after it, *Electra Glide in Blue* (1973).

From 1967, all Harley's lightweights were Italian-built, leaving the Milwaukee factory to turn out Sportsters, Electra Glides, and the Servi-Car. But foreign competition was taking significant chunks out of Harley's market share and production figures were in decline. Despite being offered on the stock market in 1965, extra cash was still needed and it was decided

•1960	1961	•1962	1963	1964	•1965	•1966	•1967	1968•	1969

FLH DUO-GLIDE

Harley buys stake in Tomahawk fiberglass company; "tombstone" speedometer on twins

ELECTRA GLIDE

74cu. in. "Shovelhead" engine introduced

CRTT

Cal Rayborn wins two Daytona 200-mile races on a Harley

twin engine and frame with the front end of a Sportster to create a new style of factory-built custom bike that is the basis of Harley's success today. And since the FX, Harley-Davidson has realized the benefits of putting out a range of models all based on similar engines, but with different styles.

Harley's lowest point

The 1970s was not a good decade for the motor industry in general and Harley in particular. The oil crisis and tooling problems resulting from the merger with AMF hit sales hard. This was despite the introduction of the XR750, which would go on to become the most successful dirt-track bike in the history of the sport. Harley bowed to the inevitable in 1978 and sold off its interest in Aermacchi, thereby ending its brief

EVEL KNIEVEL PREPARES FOR ANOTHER DEATH-DEFYING STUNT ON A HARLEY-DAVIDSON XR750

THE ALL-CONQUERING HARLEY-DAVIDSON XR750

that Harley-Davidson needed a heavyweight partner. In January 1969, AMF (American Metal Foundries) bought a controlling stake in the company.

By this time, the customizing trend had switched from "Bobbers" to "Choppers." Bikes were given raked frames and improbably long forks, wild paint jobs, and decorative chrome. These additions looked amazing, but often affected performance and handling. Harley-Davidson officially frowned on this trend, but the director of styling, William G. Davidson—the grandson of founder William A. Davidson—was watching it with interest. In 1971 he paid homage to the chopper craze—and to the look popularized by the film *Easy Rider* (1969)—with the FX1200 Super Glide. This new model combined the big-

flirtation with lightweights. By 1979 Harley sales made up just four-percent of the US market and its bikes were seen as unreliable and idiosyncratic. As the company moved into the next decade it was in desperate need of better quality control and a broader range of products. Things had to improve before the bikes appealed to a wider range of buyers.

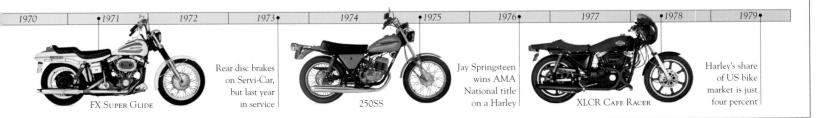

1970 •1971 1972 1973• 1974 •1975 1976• 1977 •1978 1979•

FX SUPER GLIDE

Rear disc brakes on Servi-Car, but last year in service

250SS

Jay Springsteen wins AMA National title on a Harley

XLCR CAFE RACER

Harley's share of US bike market is just four percent

1980 TO 2003

In 1980 Harley-Davidson was in big trouble. Its market share was small, reliability troubles meant its reputation was in tatters, and its machines were hopelessly outdated compared to the Japanese opposition. People who learned to ride on trouble-free Japanese machines might have liked the idea of buying a Harley, but found the reality unacceptable.

Noise and emission restrictions were also hitting Harley's old-fashioned engines hard; they got less powerful and more emasculated with each new law that was passed. Even US police departments, which had first bought bikes from Harley-Davidson in 1907 and had continued ever since, were now deserting the company and switching to foreign bikes. What Harley-Davidson needed was to introduce new machines that maintained the essence of the American V-twin, but came with vastly improved performance and reliability.

The FLT Tour Glide of 1980 was a good start. A new frame, which isolated the engine and reduced vibration, made riding comfort superior, and the revised design and geometry also improved handling—as did the frame-mounted fairing and Harley's first five-speed transmission. But nomatter how good the model was, introducing new bikes was only dealing with the problem at a purely superficial level—Harley's troubles ran much deeper than that. A year later thirteen Harley-Davidson executives, including Willie G. Davidson, bought the company back from AMF. Operating as an independent company again, it was the start of a new era.

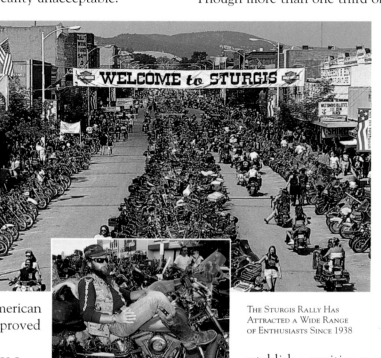

THE STURGIS RALLY HAS ATTRACTED A WIDE RANGE OF ENTHUSIASTS SINCE 1938

Big changes followed almost immediately, and they had to, considering that in 1981 the US market was being flooded with more cheap Japanese imports than at any time in its history. An efficiency drive resulted in the introduction of materials-as-needed production techniques, whereby components were delivered just prior to machine assembly. Though more than one third of the workforce was laid off as a result, the new procedure put an end to the inefficient practice of maintaining stagnant stock. Quality control was also improved and the new model development program was shifted up two gears. Harley-Davidson petitioned the International Trade Commission (ITC) for tariff restrictions on Japanese motorcycles to give it time to turn the company around and in 1982 President Ronald Reagan duly obliged by slapping tariffs on all Japanese machines over 700cc. As the new management team sought to establish a positive new direction for the company, Harley-Davidson also began an aggressive campaign to protect its trademarks and copyright.

The self-belief returns

A couple of contributory factors around this time helped Harley's recovery. The Reagan years marked a return of America's patriotism and self-belief, and the Harley-Davidson motorcycle was the ideal representation of that belief. In addition, buyers had started to doubt the value of continually searching for increased performance. They were now looking to buy bikes that made them feel good, rather

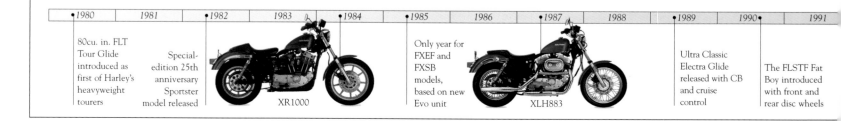

•1980	1981	•1982	1983	•1984	•1985	1986	•1987	1988	•1989	1990•	1991
80cu. in. FLT Tour Glide introduced as first of Harley's heavyweight tourers	Special-edition 25th anniversary Sportster model released		XR1000		Only year for FXEF and FXSB models, based on new Evo unit		XLH883		Ultra Classic Electra Glide released with CB and cruise control		The FLSTF Fat Boy introduced with front and rear disc wheels

HARLEY-DAVIDSON ON FILM: ARNOLD SCHWARZENEGGER IN *TERMINATOR II* (1991)

For Harley-Davidson, it was an astonishing recovery. In technological terms, the introduction of the Evolution big-twin engine in 1984 was the start of the new beginning. An engine which looked very similar to the Shovelhead, but which was cheaper to make, more reliable, quieter, and more powerful turned out to be the key to success. Even police departments now started buying Harleys again. And Harley-Davidson's innovative product development through the end of the 1980s and into the '90s assured the bike-buying public that it was well and truly back on track. Harley has acquired a unique understanding of what its customers want and the appearance of the radical new V-Rod in the new millennium suggests that the company is now entering a new phase in its long history.

A SAMPLE OF HARLEY-DAVIDSON'S EXTENSIVE MERCHANDISE

The Harley family

One of the things that customers want is to feel like they belong. Look at the formation of HOG (Harley Owners' Group) in 1983, a factory sponsored club which made new riders feel at home. It was a pastiche of the traditional bike club, with leathers and sew-on patches, but without the oily fingernails or the bad-ass attitude. No other enthusiast group sponsored by a manufacturer can boast over 400,000 members worldwide. Rallies such as Daytona and Sturgis attract Harley riders in their tens of thousands each year, but you don't have to own a Harley to feel this sense of belonging. Such has been the success of the company that you can now use your Harley-Davidson credit card to buy Harley after-shave, beer, or a Barbie™ doll. The Harley-Davidson Motor Company really has come a long way.

than ones that outperformed the opposition. Harley-Davidson, the only large-scale American motorcycle manufacturer, with 75 years of heritage, and products that looked like they were history in the 1950s, was perfectly poised to take advantage of the situation. And it did.

Japanese manufacturers realized the way this section of the market was going and, by the end of the 1980s, each of the four big companies had responded by offering variations on Harley's V-twin theme. But it was too late. By 1986 Harley had leapfrogged Honda to become the best-selling superheavyweight bike manufacturer in the United States, aided by one vital ingredient—heritage.

A 1920s 10 CENT STAMP FEATURING HARLEY-DAVIDSON'S POSTAL SERVICE MOTORCYCLE

1990s BELT BUCKLE COMMEMORATING 1920s STAMP

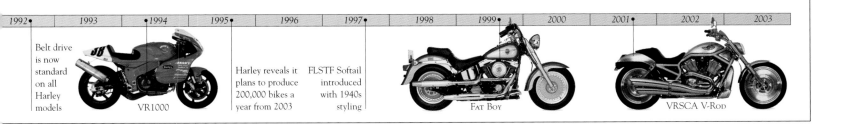

1992 • 1993 • 1994 1995 • 1996 1997 • 1998 1999 • 2000 2001 • 2002 2003

Belt drive is now standard on all Harley models

VR1000

Harley reveals it plans to produce 200,000 bikes a year from 2003

FLSTF Softail introduced with 1940s styling

FAT BOY

VRSCA V-ROD

2004 TO 2013

Harley Davidson appeared to start their second century confidently, though they were struggling with a falling share price and were soon to suffer badly in the global recession. They strived to make their bikes more efficient with better dynamics, improving performance and handling while also meeting ever more stringent emissions controls. They expanded their accessories and clothing ranges and pushed the Harley brand and products into new markets around the globe.

As well as factories at Milwaukee, the company had a substantial plant at York, Pennsylvania, assembling the big-twins, and at Kansas City, Missouri, building Sportster and V-Rod models.

Bespoke bikes

Recognizing a growing trend for bespoke custom bikes, for which well-heeled buyers would pay a premium, Harley-Davidson began offering a factory customizing service, and also a limited number of premium-price CVO (Custom Vehicle Operations) built machines with special finishes and factory-fitted accessories. From 2009 they started offering a trike too.

The importance of the brand's heritage was underscored in 2008 with the opening of a 20-acre museum in Milwaukee which contains over 450 motorcycles, and attracts hundreds of thousands of visitors every year.

Harley understood that their loyal customers were wary of them diluting the

HILL-CLIMBER SCULPTURE, HARLEY-DAVIDSON MUSEUM, MILWAUKEE

brand by producing more conventional sporting motorcycles, and, equally, customers wanting a mainstream motorcycle were suspicious of Harley-Davidsons. Harley went on to invest in Buell (with an all-new engine in 2007) and in 2008 Harley bought the famous Italian MV Agusta marque.

There was an irony here: the factory in which MVs were made had been owned by Harley-Davidson before. It was the same one in which the Harley-Davidson two-strokes had been made until 1978.

Unfortunately the global recession affected the company badly. Suddenly their market shrank rapidly and dramatic restructuring and cost-cutting were required. The Buell division was closed (just as they won their first AMA Pro Racing championship) and, less than two years after it was acquired, MV Agusta was sold at a substantial loss. Production of the company's traditional bikes was reduced and so was the workforce.

In adversity the company focused on making its existing machines better with incremental improvements in quality and innovation. They sought out new buyers, working to attract women and younger riders at home, and moved into growing markets too; they established an Indian subsidiary in 2009 and were assembling five models there by 2012.

A stuttering world economy has stalled the growth of the Harley market and reduced the company's profitability, but the bikes themselves are better than ever, combining improved efficiency with a nod to history and heritage.

AN ASSEMBLY LINE AT A HARLEY-DAVIDSON PLANT

▶ POLISHED TO A SHINE: HARLEYS PARKED IN NORTH YORKSHIRE, BRITAIN, IN 2006

2004	2005	2006	2007	2008	2009	2010	2011	2012	2013
New chassis and updated engine for Sportster range		35th anniversary Super Glide models	50th anniversary Sportster	New Rocker Softail models	XR1200X		XL883N Iron		Company celebrates 110 years

CHAPTER ONE
THE EARLY BIKES
1903–1929

1905 SINGLE CYLINDER

EARLY INTERNAL COMBUSTION ENGINES were underpowered, unreliable, and almost comically crude. Part of Harley-Davidson's success was its realization that increasing engine size was part of the solution to lack of power and that practical design and added strength improved reliability. But it didn't embrace change for change's sake. The evolution of the Harley engine from the first prototypes to the last of the inlet-over-exhaust valve machines of 1929 was gradual and evolutionary.

EARLY BIKE BROTHER

Walter Davidson stands next to an early Harley single (left). After building their first bike in 1903, it took three years before the four founders worked full-time at Harley-Davidson.

1905 Model No.1

MOTOR AND CYCLE. There wasn't much more to the first Harleys than these two vital ingredients. Though William Harley and the Davidson brothers used a larger engine than most of their contemporary manufacturers, the fact that pedal power was an essential supplement to the internal combustion engine on hills meant that the bicycle layout had to be retained. Harley curved the bottom frame tube under the crankcases to allow the engine to be mounted lower in the frame, resulting in superior handling. The battery ignition system, crude carburetor, belt drive, and other unrefined elements ensured that the early motorcycle wasn't really a viable means of transportation, but at least the Model No.1 was a cut above the average.

Oil tank mounted within fuel tank

Rod linkage connects twist-grip control to carburetor

Carburetor float chamber

White rubber tires were common on early machines until it was realized that black rubber hid the dirt

Solid bicycle-style front forks

One-piece cast-iron cylinder

Drive pulley

66-in (26-cm) wheel

1905 MODEL NO.1

The 1905 Model No.1 was almost identical to the bikes built in 1903 and '04, and until 1909 Harley-Davidson produced only one model, which was improved upon each year. The model number represented the year of production minus four.

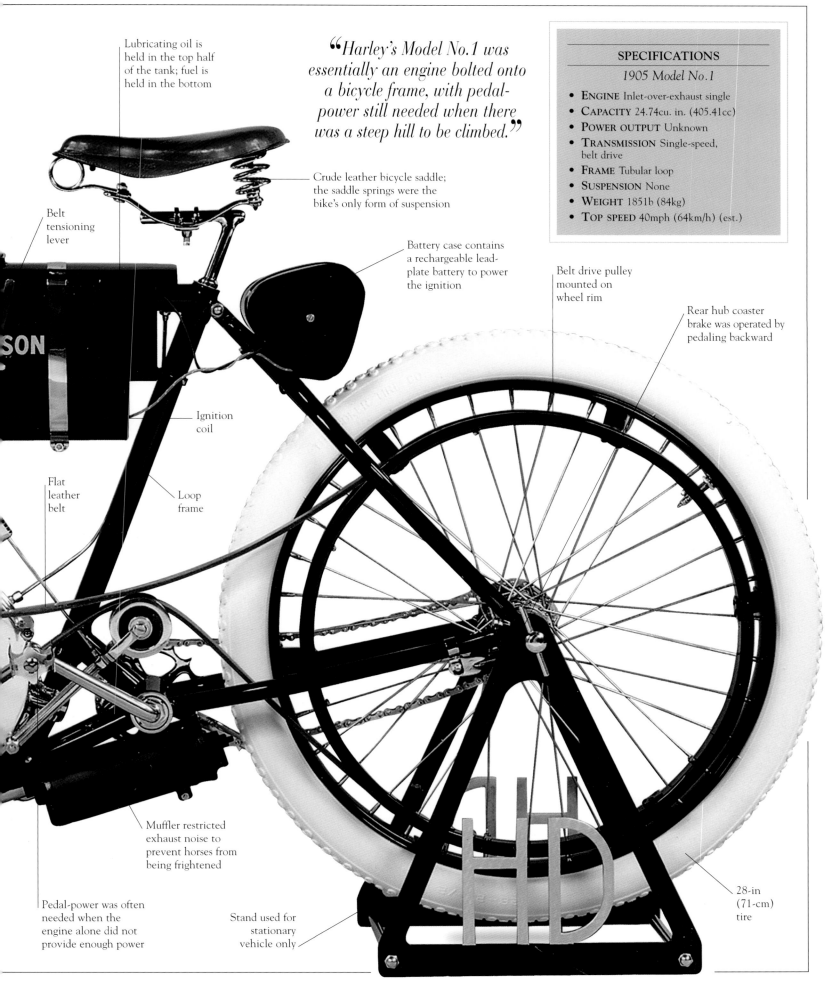

Lubricating oil is held in the top half of the tank; fuel is held in the bottom

"Harley's Model No.1 was essentially an engine bolted onto a bicycle frame, with pedal-power still needed when there was a steep hill to be climbed."

Crude leather bicycle saddle; the saddle springs were the bike's only form of suspension

Belt tensioning lever

Battery case contains a rechargeable lead-plate battery to power the ignition

Belt drive pulley mounted on wheel rim

Rear hub coaster brake was operated by pedaling backward

Ignition coil

Flat leather belt

Loop frame

Muffler restricted exhaust noise to prevent horses from being frightened

Pedal-power was often needed when the engine alone did not provide enough power

Stand used for stationary vehicle only

28-in (71-cm) tire

SPECIFICATIONS
1905 Model No.1

- **ENGINE** Inlet-over-exhaust single
- **CAPACITY** 24.74cu. in. (405.41cc)
- **POWER OUTPUT** Unknown
- **TRANSMISSION** Single-speed, belt drive
- **FRAME** Tubular loop
- **SUSPENSION** None
- **WEIGHT** 185lb (84kg)
- **TOP SPEED** 40mph (64km/h) (est.)

1912 Silent Gray Fellow

BY THE TIME HARLEY-DAVIDSON built this X-8 single in 1912, the company was well on the way to establishing itself as a major motorcycle manufacturer, and the motorcycle was a more refined mode of transportation. The rugged engineering and rigorous development championed by Harley from day one had borne fruit in the form of sprung forks and magneto ignition, and the company wasted no time emphasizing that cubic inches were the key to increased power. The original 1903 Harley had a 24.74cu. in. (405cc) engine, rising to 26.8cu. in. (440cc) in 1906, and 30cu. in. (494cc) in 1909. In 1913 it gained a further 5cu. in. (82cc). The Harley single became a valued and dependable machine which earned it the nickname "Silent Gray Fellow."

1912 SILENT GRAY FELLOW
This bike was a direct development of the original 1903 model and continued in production until 1918. Though it still had belt final-drive, an atmospheric inlet valve, and no gearbox, these developments were just around the corner.

Atmospheric inlet valve is kept closed by this light spring

Control cables replaced rod linkages in 1909

Valanced front fender

White rubber tires were a period feature

Leading-link front suspension

Canvas mudflap

Loop frame curves under the engine, allowing it to be positioned lower for optimum weight distribution

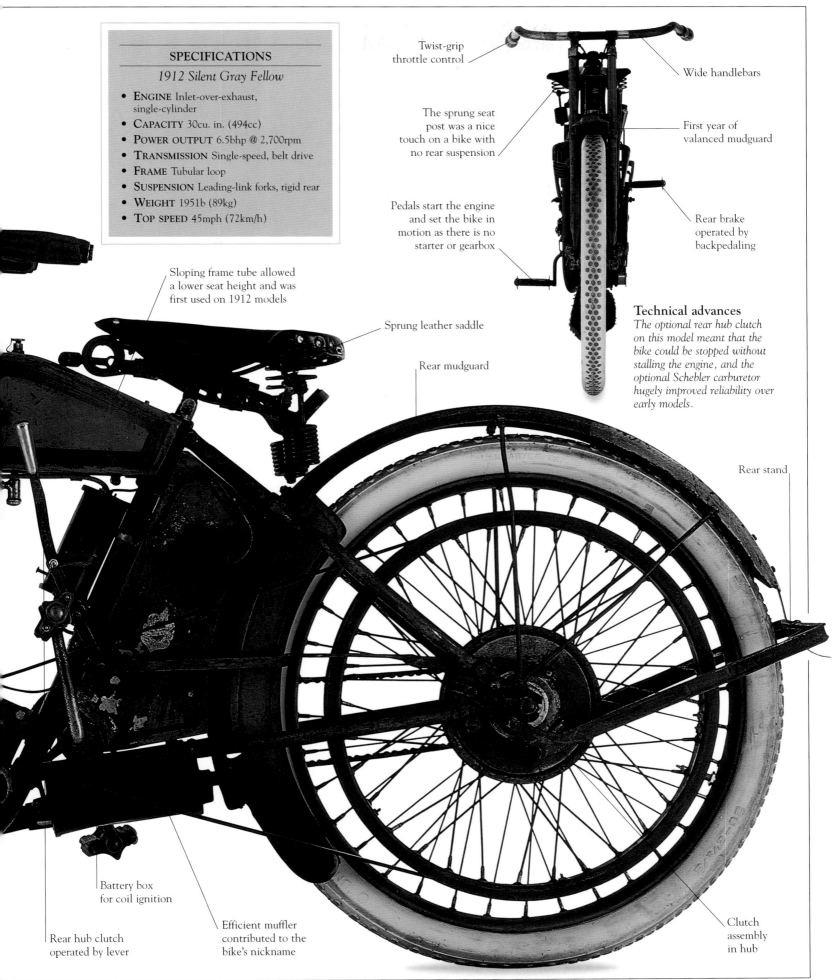

SPECIFICATIONS

1912 Silent Gray Fellow

- **ENGINE** Inlet-over-exhaust, single-cylinder
- **CAPACITY** 30cu. in. (494cc)
- **POWER OUTPUT** 6.5bhp @ 2,700rpm
- **TRANSMISSION** Single-speed, belt drive
- **FRAME** Tubular loop
- **SUSPENSION** Leading-link forks, rigid rear
- **WEIGHT** 195lb (89kg)
- **TOP SPEED** 45mph (72km/h)

Twist-grip throttle control

Wide handlebars

The sprung seat post was a nice touch on a bike with no rear suspension

First year of valanced mudguard

Pedals start the engine and set the bike in motion as there is no starter or gearbox

Rear brake operated by backpedaling

Sloping frame tube allowed a lower seat height and was first used on 1912 models

Sprung leather saddle

Rear mudguard

Technical advances
The optional rear hub clutch on this model meant that the bike could be stopped without stalling the engine, and the optional Schebler carburetor hugely improved reliability over early models.

Rear stand

Battery box for coil ignition

Efficient muffler contributed to the bike's nickname

Clutch assembly in hub

Rear hub clutch operated by lever

The Early Single

THE EARLY SINGLE-CYLINDER gasoline engine was underpowered and inefficient, though Harley's singles were among the best available. The "atmospheric" inlet valve relied on a light spring to keep it closed and was forced open by the pressure created from the falling piston. The system was simple, but it couldn't work properly at anything other than slow engine speeds. Increasing capacity boosted the power output, but it was no substitute for improved efficiency.

Single with Bosch magneto
A magneto was introduced as an option in 1909. This simple electric generator provided a spark for the ignition system and made the early Harley a more useable machine.

Throttle control linkage

Lead

High-tension lead

Spark plug

Iron barrel and cylinder head were cast as one piece

Atmospheric inlet valve stem and spring

Vertical fins on the cylinder head were introduced in 1911 to aid engine cooling

Carburetor throat

Harley introduced the Schebler float-feed carburetor on its bikes from 1909; the company had previously made its own

Bosch high-tension magneto was first seen on Harley singles in 1909

Exhaust valve stem and spring

No female frills
Some European manufacturers offered "ladies' models" with special frames and skirt-guards. Harley women didn't need these luxuries.

Oil feed union; a pipe supplied lubricant to the crankshaft, with pressure created by the hand-operated pump in the tank

INLET-OVER-EXHAUST
The inlet valve was kept closed with the exposed spring on the top of the cylinder. As the piston fell, it created a vacuum in the cylinder, resulting in the valve being forced open by atmospheric pressure. A charge of fuel and air mix was then sucked in through the carburetor.

The exhaust pipe was connected to a simple muffler, which could be bypassed to increase performance and noise

The timing gear case conceals the four gears that drove the magneto; models with battery ignition were not equipped with gears

Engine case bolt

Alloy engine case bears the legend "Harley-Davidson, Milwaukee"; the city's other notable product was, and is, beer

Alloy crankcase

THE COMPETITION
• 1911 EXCELSIOR MODEL K •
The Chicago Excelsior company was the third-biggest motorcycle manufacturer in the US until its closure in 1931. Like the Harley, this machine had belt-drive, but used the engine cases as part of the frame, as opposed to Harley's loop-frame system.

❝The Harley single was just a good, solid, dependable motorcycle at a time when most of them were not.❞

RICHARD ROSENTHAL
(MOTORCYCLE HISTORIAN)

Rugged and dependable
In an era of unreliable and uncomfortable motorcycles, the Harley single stood out as a solid workhorse capable of covering long distances. The well-made engine got some of the credit, but Harley's sprung seat-post was a welcome feature for the often appalling road conditions.

1900s | 1910s | 1920s | 1930s | 1940s | 1950s | 1960s | 1970s | 1980s | 1990s | 2000s

1915 KT Board Racer

HARLEY-DAVIDSON DID NOT participate in team racing until 1914, when it decided to exploit the potential benefits of publicity and development that could be derived from racing success. Board-track racing was reaching new levels of popularity, with promotors able to attract huge paying crowds to the meetings, so Harley's decision to enter into competition made a lot of sense. And the move paid off almost immediately, as the Harley race team began to achieve significant results in 1915 on bikes such as this KT. In September 1915, an F-head Harley set a 100-mile (161-km) record of 89.11mph (143.46km/h) on a board track in Chicago. It all augured well for the launch of the famous eight-valve racer (*see pp.46–47*) a year later.

SPECIFICATIONS
1915 KT Board Racer

- **ENGINE** Inlet-over-exhaust, V-twin
- **CAPACITY** 61cu. in. (1000cc)
- **POWER OUTPUT** 15bhp
- **TRANSMISSION** Three-speed, chain drive
- **FRAME** Tubular loop
- **SUSPENSION** Leading-link front forks, rigid rear
- **WEIGHT** 325lb (147kg)
- **TOP SPEED** 80mph (130km/h)

Fuel filler-cap

Inlet-valve pushrod

Basic lightweight saddle provided little comfort

Abbreviated mudguard is one of a number of weight-saving components

The brake on this bike is a luxury; most board racers didn't have one

Magneto; the KTH version came with an electrical system instead

Pedaling forward starts the bike, pedaling backward operates the rear brake—just like on an American bicycle

Window in the timing gear case allowed the rider to see if the oil pump was working

Lightweight wheel hub

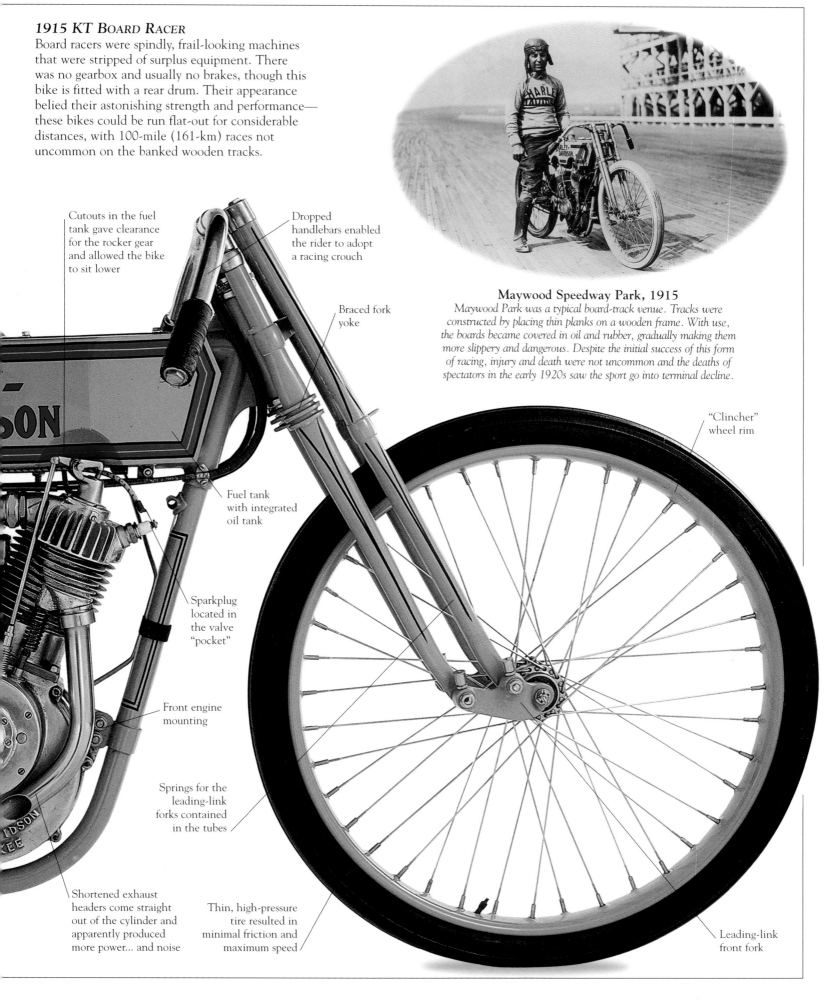

1915 KT BOARD RACER

Board racers were spindly, frail-looking machines that were stripped of surplus equipment. There was no gearbox and usually no brakes, though this bike is fitted with a rear drum. Their appearance belied their astonishing strength and performance—these bikes could be run flat-out for considerable distances, with 100-mile (161-km) races not uncommon on the banked wooden tracks.

Cutouts in the fuel tank gave clearance for the rocker gear and allowed the bike to sit lower

Dropped handlebars enabled the rider to adopt a racing crouch

Braced fork yoke

Maywood Speedway Park, 1915
Maywood Park was a typical board-track venue. Tracks were constructed by placing thin planks on a wooden frame. With use, the boards became covered in oil and rubber, gradually making them more slippery and dangerous. Despite the initial success of this form of racing, injury and death were not uncommon and the deaths of spectators in the early 1920s saw the sport go into terminal decline.

"Clincher" wheel rim

Fuel tank with integrated oil tank

Sparkplug located in the valve "pocket"

Front engine mounting

Springs for the leading-link forks contained in the tubes

Shortened exhaust headers come straight out of the cylinder and apparently produced more power... and noise

Thin, high-pressure tire resulted in minimal friction and maximum speed

Leading-link front fork

1915 KR Fast Roadster

THE IDEA OF A RACE BIKE on the road has always been attractive to motorcyclists. Modern bikers relish the power, handling, and brakes of competition-developed machinery and pioneer motorcyclists were no different. Harley's Fast Roadster was based on the board-track racer (*see pp.32–33*) but fitted with mudguards, a chainguard, and conventional handlebars. Who needed a gearbox or lights? It was built for amateur racers at a time when Harley-Davidson's factory race team was starting to taste success; the K-series won a number of 100- and 300-mile (161- and 483-km) races for Harley in 1915. With just over 100 built, this model is now very rare.

1915 KR FAST ROADSTER
The "close coupled" frame on the KR was shorter than that fitted to other models such as the F (*see pp.38–39*) because no gearbox was fitted. A short wheelbase traditionally offers better handling at the expense of comfort and stability.

The metal tank was made up of three reservoirs; two contained fuel and one on the front-left held oil

Hand oil pump

Control cables made of single-strand piano wire with a wound metal outer cable, leather-sheathed for protection

"Clincher" wheel rims were used with beaded tires which required very high pressures to stay on the rims

Clutch pedal

Loop frame

Leading-link front fork

Canvas mudguard

The Fast Roadster had sufficient power to make the bicycle-style pedals on earlier bikes unnecessary

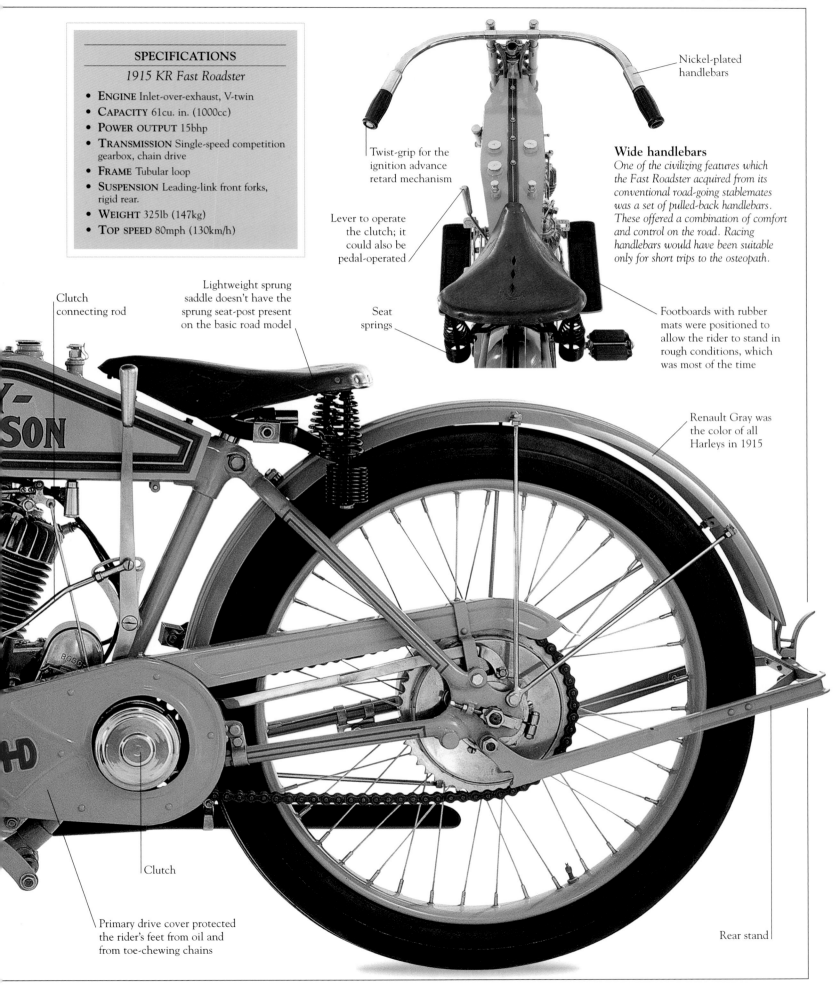

SPECIFICATIONS

1915 KR Fast Roadster

- **ENGINE** Inlet-over-exhaust, V-twin
- **CAPACITY** 61cu. in. (1000cc)
- **POWER OUTPUT** 15bhp
- **TRANSMISSION** Single-speed competition gearbox, chain drive
- **FRAME** Tubular loop
- **SUSPENSION** Leading-link front forks, rigid rear.
- **WEIGHT** 325lb (147kg)
- **TOP SPEED** 80mph (130km/h)

Twist-grip for the ignition advance retard mechanism

Lever to operate the clutch; it could also be pedal-operated

Nickel-plated handlebars

Wide handlebars

One of the civilizing features which the Fast Roadster acquired from its conventional road-going stablemates was a set of pulled-back handlebars. These offered a combination of comfort and control on the road. Racing handlebars would have been suitable only for short trips to the osteopath.

Seat springs

Footboards with rubber mats were positioned to allow the rider to stand in rough conditions, which was most of the time

Clutch connecting rod

Lightweight sprung saddle doesn't have the sprung seat-post present on the basic road model

Renault Gray was the color of all Harleys in 1915

Clutch

Primary drive cover protected the rider's feet from oil and from toe-chewing chains

Rear stand

The 45° F-Head V-Twin

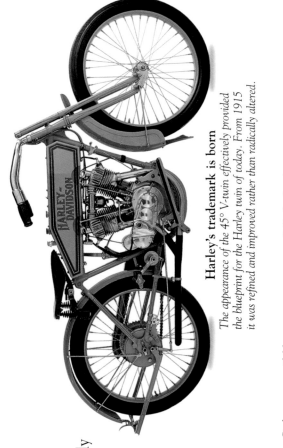

Harley's trademark is born
The appearance of the 45° V-twin effectively provided the blueprint for the Harley twin of today. From 1915 it was refined and improved rather than radically altered.

HARLEY-DAVIDSON SOON realized that the easiest way to significantly increase engine size was to add an extra cylinder. The company built a prototype V-twin in 1907 and four years later the first production V-twins rolled out of the factory. The engine effectively joined two singles on a common crankshaft and cases. If the angle of the "V" was narrow, the new engine could be used in the same frame as a single. Harley chose a 45° "V", and a motorcycling classic was born. The arrival of the mechanical exhaust valve on the V-twin was also important, allowing engine revs to be increased and thus release more power.

Thread for rocker arm assembly

Fins to aid cooling

Inlet and exhaust valves face each other in the cylinder "pocket"

Cylinder head was cast together with the barrel and so could not be removed

Carburetor manifold connects the two inlet ports and is the ideal setup for a single carburetor

Exhaust valve stem

Inlet pushrod fits to the rocker

Adjustment sleeve

Rocker arm assembly for the inlet valve fits to a threaded insert in the cylinder head

Inlet valve cage

Cast-iron was used for the cylinders because of its heat dissipation qualities and its resistance to wear

Threaded exhaust port exits beneath the exhaust valve

CHAPTER TWO

EARLY INNOVATIONS

1918–1942

1920 EIGHT-VALVE RACER

HARLEY'S FIRST ATTEMPT to produce something different from the V-twin was a 1919 fore-and-aft lightweight flat-twin, followed by utilitarian singles and a number of innovative competition racers. The shaft-drive military XA rounded off a period when, despite the presence of the V-twin, Harley-Davidson was still developing innovative new bikes.

1925 WINNING RACE TEAM
Harley's factory race team secured numerous victories on eight-valve and two-cam racers, bikes that were technologically well ahead of the competition at the time.

1918 Model J Sidecar

Harley-Davidson FIRST ADDED sidecars to its model line in 1914, and later offered specially tuned engines for sidecar use. Before then, standard bikes such as this Model J just had a sidecar bolted onto them. A sidecar meant motorcycle riders could now transport their family, large packages, or even a nervous crinolene-clad girlfriend, for whom the sidecar was a poor substitute for a proper motor car. The sidecar peaked in the years leading up to the 1920s, with some Harley examples even used on the battlefields of World War I, but the Model T Ford made the car cheaper, and from around 1920 the sidecar became a minority interest for the eccentric enthusiast.

1918 MODEL J SIDECAR
Harley's big F-head V-twin was ideal for pulling a sidecar and the company began offering sidecars as an option in 1914. From then until 1925 Harley's sidecars were built by the Rogers Company, but when Rogers ceased production Harley started building its own chairs. Production has continued, but they are now built in small numbers.

Hand-operated horn

Acetylene lighting was a period addition for bikes not equipped with electric lighting at the factory

Olive Green paint scheme was standard for 1918

Valanced front mudguard

Whitewall tire

Leading-link front suspension

61cu. in. F-head engine was offered with a special sidecar tune

Foot pedal for clutch

Hand shift lever for three-speed gearbox

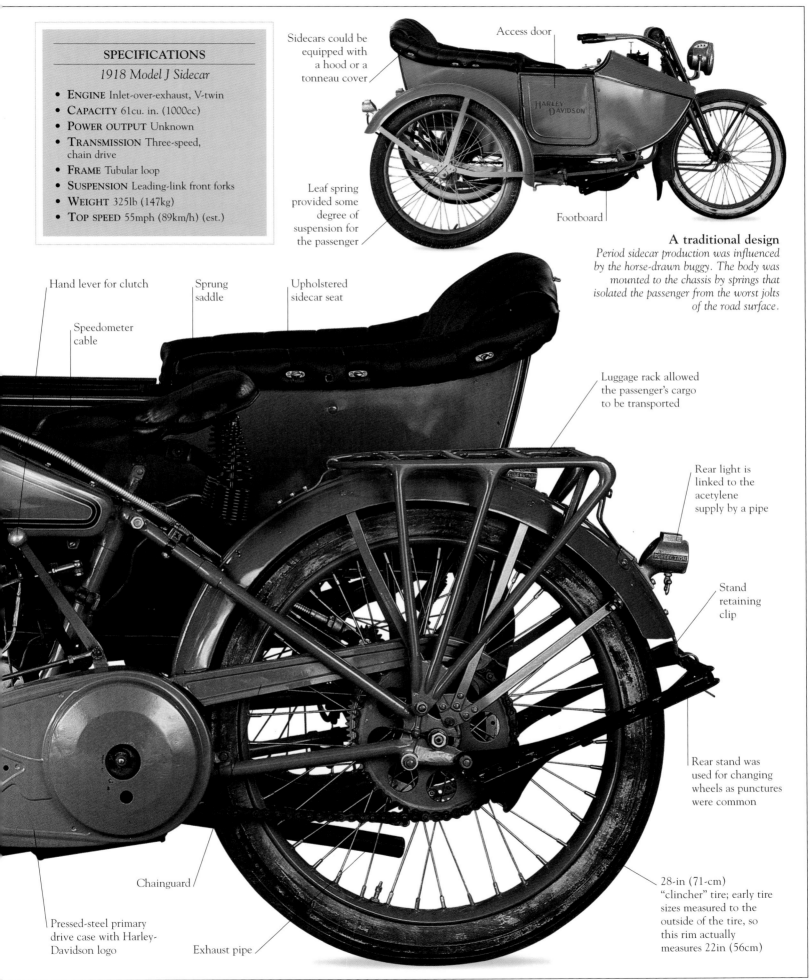

SPECIFICATIONS
1918 Model J Sidecar

- **ENGINE** Inlet-over-exhaust, V-twin
- **CAPACITY** 61cu. in. (1000cc)
- **POWER OUTPUT** Unknown
- **TRANSMISSION** Three-speed, chain drive
- **FRAME** Tubular loop
- **SUSPENSION** Leading-link front forks
- **WEIGHT** 325lb (147kg)
- **TOP SPEED** 55mph (89km/h) (est.)

Access door

Sidecars could be equipped with a hood or a tonneau cover

Leaf spring provided some degree of suspension for the passenger

Footboard

A traditional design
Period sidecar production was influenced by the horse-drawn buggy. The body was mounted to the chassis by springs that isolated the passenger from the worst jolts of the road surface.

Hand lever for clutch

Speedometer cable

Sprung saddle

Upholstered sidecar seat

Luggage rack allowed the passenger's cargo to be transported

Rear light is linked to the acetylene supply by a pipe

Stand retaining clip

Rear stand was used for changing wheels as punctures were common

Chainguard

Pressed-steel primary drive case with Harley-Davidson logo

Exhaust pipe

28-in (71-cm) "clincher" tire; early tire sizes measured to the outside of the tire, so this rim actually measures 22in (56cm)

1920 Eight-Valve Racer

HAVING COMMITTED ITSELF TO bike racing in 1914, Harley soon began to take the sport seriously. Special eight-valve racing twins were introduced in 1916; These were built in very limited numbers until 1927 for the use of the factory's own race team and chosen riders. Four versions of the machine were produced over an 11-year period, giving serious credibility to Harley as a racing-bike manufacturer. The race team secured numerous victories on the eight-valve racers and earned itself the nickname "the Wrecking Crew." The sight and sound of these—quite literally—fire-breathing machines must have been incredible as they reached speeds of around 120mph (193km/h) on the steeply banked wooden tracks.

1920 EIGHT-VALVE RACER
The cylinders and heads on the eight-valve racers were the work of British engineer Harry Ricardo and featured a hemispherical combustion chamber that had been developed on airplane engines during WW1. The bike shown here is a 1920 version with distinct open-port cylinder heads that have no exhaust headers. It is possibly one of only eight built, with very few of these still in existence. On the rare occasions when they are sold, they command substantial prices.

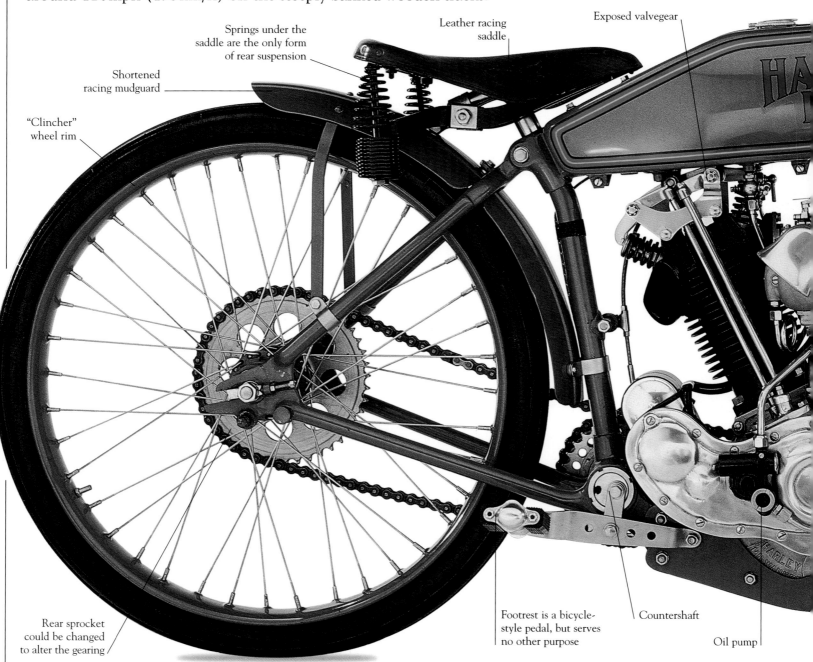

Springs under the saddle are the only form of rear suspension

Leather racing saddle

Exposed valvegear

Shortened racing mudguard

"Clincher" wheel rim

Rear sprocket could be changed to alter the gearing

Footrest is a bicycle-style pedal, but serves no other purpose

Countershaft

Oil pump

SPECIFICATIONS
1920 Eight-Valve Racer

- **ENGINE** Overhead-valve, V-twin
- **CAPACITY** 61cu. in. (1000cc)
- **POWER OUTPUT** 15bhp
- **TRANSMISSION** Single-speed, direct drive
- **FRAME** Tubular loop Keystone racing frame
- **SUSPENSION** Leading-link front forks, rigid rear
- **WEIGHT** 692lb (314kg)
- **TOP SPEED** 120mph (193km/h)

Twist-grip throttle

Fork spring tube

Carburetor air intake

Dropped handlebars forced the rider into a racing crouch

Friction damper makes suspension movement more controllable

Left twist-grip controls the ignition advance retard

Footrest

(You've got to) brake-free
The eight-valve racer had no gearbox or brakes. Riders slowed their machines using a combination of the throttle, the engine-kill button, and old-fashioned boot leather. Harley riders, however, were up to the task, winning a number of prestigious races this year.

Throttle cable runs inside the handlebars

Spoked racing wheel

Compression-release lever

Open exhaust port

Oil feed pipe to the front cylinder; crank rotation forced oil back to the rear cylinder

Compression-release mechanism can be used to kill the engine or to allow the clutchless bike to be pushed with a dead engine

Engine mounting plate

Suspension linkage

Thin high-pressure tire for minimum friction

The Eight-Valve

TWO INLET VALVES AND TWO EXHAUST valves per cylinder allowed gases to flow into and out of the engine in greater quantity than in the traditional four-valve V-twin. The resulting increase in power was exactly what was needed for a successful racing engine to be used in Harley-Davidson's newly formed factory race team. Harley wasn't the first manufacturer to use this technology, but the company's eight-valve racers carved a massive reputation based on numerous racing victories.

Single speed but extra power
This profile shows the gearing on the eight-valve racer. This version has no clutch, gearbox, or brakes, and could exceed 100mph (161km/h).

Rocker arm

Open exhaust port followed period aircraft engine style, though some eight-valves were fitted with short pipes

Cast-iron cylinders each have a capacity of 30.50 cu. in., to give a total capacity of 61cu. in.

Throttle control linkage

Air intake trumpet for single carburetor

Gasoline tap connects to fuel tank

Oil feed to front cylinder

Rear cylinder exhaust valve and spring

External pushrod

Carburetor

Bosch magneto

Exposed valvegear

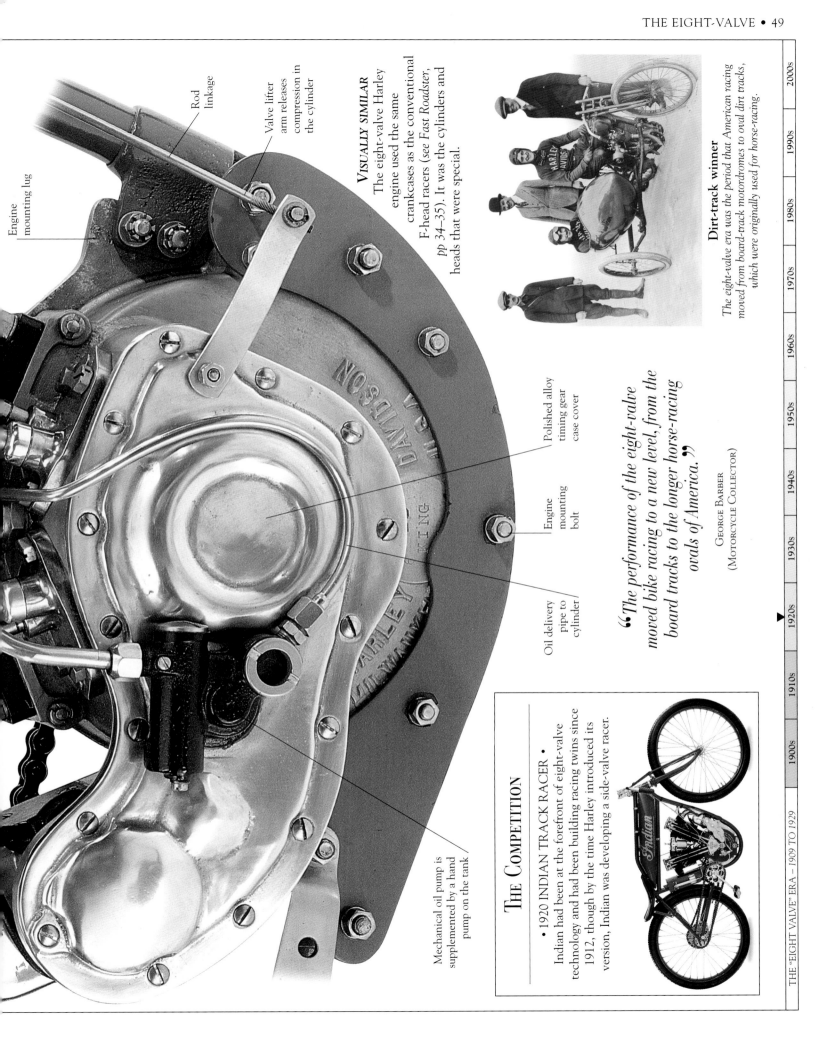

Engine mounting lug

Rod linkage

Valve lifter arm releases compression in the cylinder

VISUALLY SIMILAR

The eight-valve Harley engine used the same crankcases as the conventional F-head racers (see *Fast Roadster*, pp 34–35). It was the cylinders and heads that were special.

Polished alloy timing gear case cover

Engine mounting bolt

Oil delivery pipe to cylinder

Mechanical oil pump is supplemented by a hand pump on the tank

Dirt-track winner
The eight-valve era was the period that American racing moved from board-track motordromes to oval dirt tracks, which were originally used for horse-racing.

> *"The performance of the eight-valve moved bike racing to a new level, from the board tracks to the longer horse-racing ovals of America."*
>
> GEORGE BARBER
> (MOTORCYCLE COLLECTOR)

THE COMPETITION

• 1920 INDIAN TRACK RACER •
Indian had been at the forefront of eight-valve technology and had been building racing twins since 1912, though by the time Harley introduced its version, Indian was developing a side-valve racer.

2000s 1990s 1980s 1970s 1960s 1950s 1940s 1930s 1920s 1910s 1900s

1926 Model B

AFTER THE COMPARATIVE FAILURE of the Sport Twin of 1919–23, Harley had another crack at the lightweight market by releasing a range of single-cylinder bikes for the 1926 model year: the A, B, AA, and BA. The design was entirely conventional, and inspired by Indian's contemporary Prince as well as typical British machines of the period. The bikes were available with side-valve or overhead-valve engines and the racing versions that followed were nicknamed "Peashooters" due to the unique pitch of the exhaust; The name was eventually applied to all the models. These were ideal machines for impoverished commuters and delivery riders who accepted underwhelming performance as long as the bike was cheap to buy and run.

SPECIFICATIONS
1926 Model B

- **ENGINE** Side-valve, single cylinder
- **CAPACITY** 21cu. in. (346cc)
- **POWER OUTPUT** 10bhp
- **TRANSMISSION** Three-speed, chain drive
- **FRAME** Tubular loop
- **SUSPENSION** Leading-link front forks, rigid rear
- **WEIGHT** 263lb (119kg)
- **TOP SPEED** 60mph (97km/h)

Olive Green with maroon striping was the standard paint finish on all Harley bikes up until 1933

26x3⅓-in (66x8.5-cm) "balloon" tire

Bicycle-style seat features Harley's traditional telescopic sprung mounting for increased comfort

Taillight; the budget-priced Model A wasn't fitted with lights

Tool roll

Rear stand; a side-stand could be fitted at extra cost

Muffler

The battery allowed the bike to run lights; buyers who chose the magneto option had no lights

Exhaust pipe

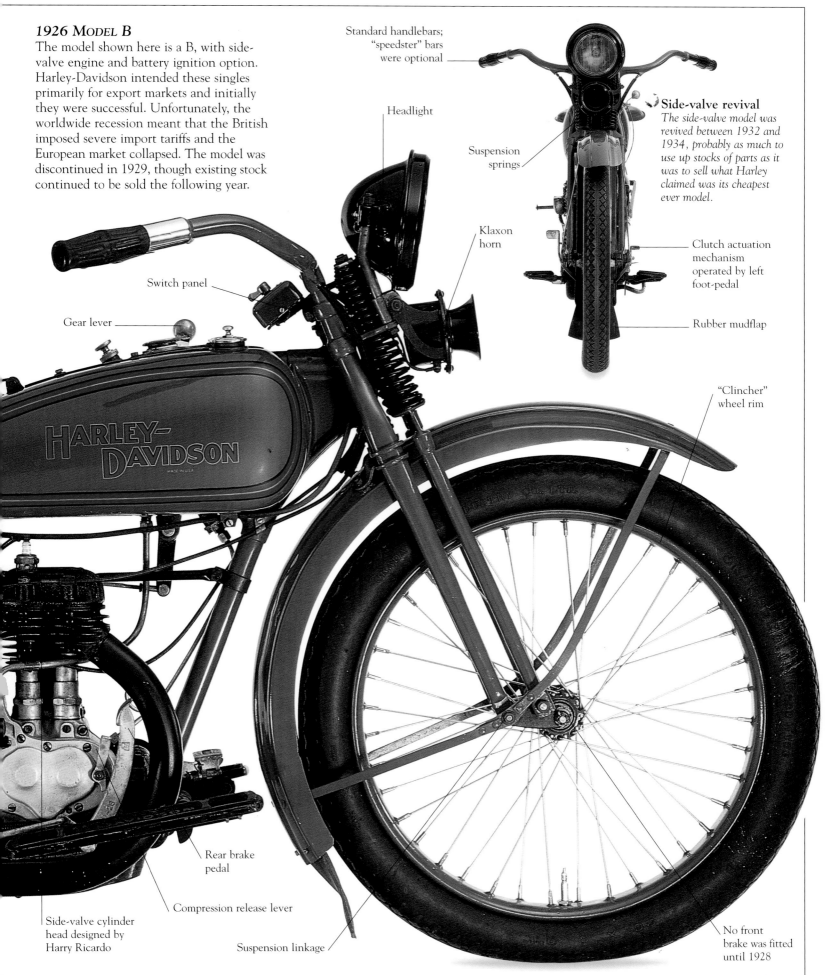

1926 MODEL B

The model shown here is a B, with side-valve engine and battery ignition option. Harley-Davidson intended these singles primarily for export markets and initially they were successful. Unfortunately, the worldwide recession meant that the British imposed severe import tariffs and the European market collapsed. The model was discontinued in 1929, though existing stock continued to be sold the following year.

Standard handlebars; "speedster" bars were optional

Headlight

Suspension springs

Side-valve revival
The side-valve model was revived between 1932 and 1934, probably as much to use up stocks of parts as it was to sell what Harley claimed was its cheapest ever model.

Klaxon horn

Clutch actuation mechanism operated by left foot-pedal

Rubber mudflap

Switch panel

Gear lever

"Clincher" wheel rim

Rear brake pedal

Compression release lever

Side-valve cylinder head designed by Harry Ricardo

Suspension linkage

No front brake was fitted until 1928

1926 Model S Racer

A NEW 350CC RACING class was created soon after Harley unveiled its "Peashooter" racer in the summer of 1925. The bike was based on its new 21cu. in. ohv single-cylinder economy road bike. To make it competitive for dirt-track racing the bike had a shortened frame and simple telescopic forks that were triangulated for greater strength. The legendary Joe Petrali was among several riders who achieved success on Peashooters as Harleys swept the board in the new class. Petrali was one of the best riders in the history of American bike racing, and in 1935 he won all 13 rounds of the US dirt-track championships on a stock Peashooter.

1926 MODEL S RACER

Harley produced its Peashooter racing bikes in limited numbers for a few years and, as well as success at home, they were also raced successfully in Britain and Australia. However, the appearance of the British JAP-powered machines in the 1930s effectively made all competitors redundant and the Peashooter disappeared from the racing circuit.

Dropped racing handlebars

Engine lubrication relies on a hand oil pump

Flimsy telescopic forks are braced for extra strength

Knobbly treaded tire was designed to provide maximum grip on dirt tracks

Exhaust outlet

Canvas fork gaiter protects the fork slider

"Clincher" wheel rims

Lefthand footrest for use on straights

Carburetor
air intake

Footrest

Racing modifications
*All nonessential components
were ditched from the standard
road model, including the
springer forks, saddle spring,
brakes, and gearbox. The bikes
ran in a single gear and the
clutch was only used for starting.
Races on oval dirt tracks were
run counter clockwise and the
left foot was used to stablilize
the bike in turns.*

SPECIFICATIONS
1926 Model S Racer

- **ENGINE** Overhead-valve,
  single cylinder
- **CAPACITY** 21cu. in. (346cc)
- **POWER OUTPUT** 12bhp
- **TRANSMISSION** Single-speed,
  chain drive
- **FRAME** Tubular loop
- **SUSPENSION** Telescopic front forks
- **WEIGHT** 240lb (109kg)
- **TOP SPEED** 70mph
  (113km/h) (est.)

Earlier versions of the
bike had a shorter,
rounded tank

Unsprung leather
saddle was not
designed for comfort

Lightweight frame

Cut-down rear
mudguard

Mudguard stay

Magneto

Footrest mounting
plate also serves as
a basic chain-guard

Drive is via primary
reduction gear and a
countershaft, with no
clutch or gearbox

Exposed primary chain drive

27-in
(68½-cm) tire

1930 Hill Climber

THE INGREDIENTS OF AN AMERICAN hill-climb bike appear simple, even if the reality is rather more complicated. The essential element is power, and in the case of this machine a methanol-burning eight-valve engine was enough in 1930 to make it a competitive bike. A long wheelbase and weight at the front to prevent the bike tipping over backward are both essential, as is grip, which is why this bike's rear tire is wrapped in chains. These crude facts belie the level of expertise involved in handling these machines, and once again it was Joe Petrali who took the honors for Harley-Davidson. Between 1932 and 1938, he won six national hill-climbing titles on a Harley.

SPECIFICATIONS
1930 Hill Climber

- **ENGINE** Eight-valve, V-twin
- **CAPACITY** 74cu. in. (1213cc)
- **POWER OUTPUT** Not available
- **TRANSMISSION** Competition single-speed gearbox
- **FRAME** Tubular cradle
- **SUSPENSION** Leading-link forks, rigid rear
- **WEIGHT** 350lb (147kg) (est.)
- **TOP SPEED** Determined by the gearing chosen for the hill-climb course

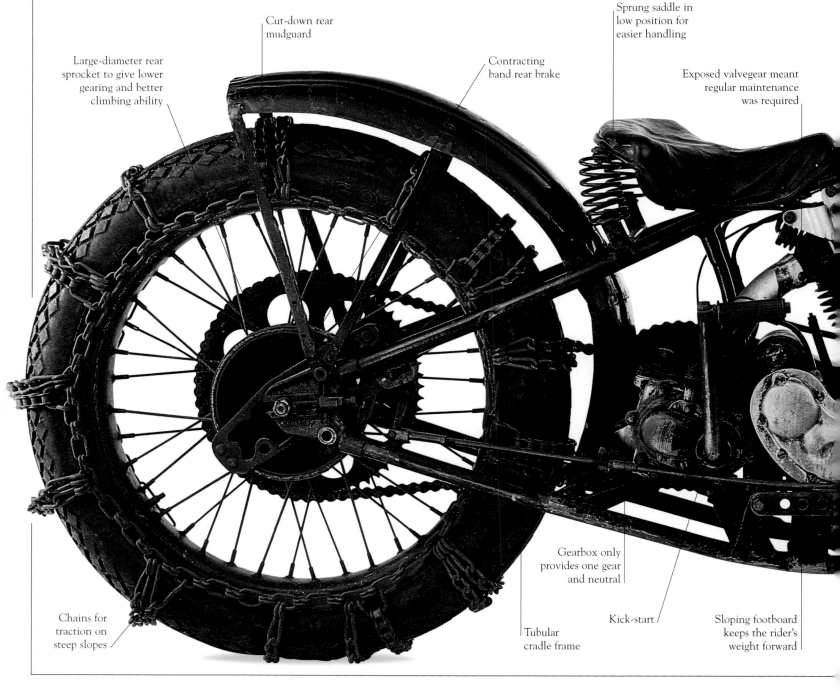

Large-diameter rear sprocket to give lower gearing and better climbing ability

Cut-down rear mudguard

Contracting band rear brake

Sprung saddle in low position for easier handling

Exposed valvegear meant regular maintenance was required

Chains for traction on steep slopes

Tubular cradle frame

Gearbox only provides one gear and neutral

Kick-start

Sloping footboard keeps the rider's weight forward

1930 HILL CLIMBER

This unrestored bike is typical of the ingenious hill-climbing machines of the period. The frame is from a JD model circa 1929, and the forks are from a 1928 45cu. in. bike. The engine cases are from a JDH (*see p.41*), with JE model flywheels; special barrels and overhead-valve cylinder heads are from a single-cylinder Harley. Modern hill-climb racers still have a similar look to this rugged machine and the difficulty of trying to convert power into climbing ability remains the same.

Racing rules

In hill-climbing, riders launch their machines at impossibly steep hills. If riders make the summit then time decides the winner, but if no rider reaches the peak, then the one who has reached the highest point wins.

Company colors

The Harley-Davidson race teams were easily recognizable by their orange and black jerseys. These colors would later be used as the livery for Harley's racing bikes.

Race jersey from the 1930s

Racing handlebars

Filler cap for small-capacity fuel tank

Schebler racing carburetor

Headstock forging is drilled to reduce weight

Heavy-duty racing wheel

Right-hand section of tank contains oil

Shortened exhaust header pipe allowed maximum power from the engine

Oil feed pipe

J-series engine was especially tuned for racing

JDH engine case has two cams

Blanked-off oil-pump drive; lubrication was by hand pump

Leading-link front fork

Ribbed front tire

1942 XA

IT IS WELL DOCUMENTED THAT Harley supplied thousands of traditional 45° V-twin WLA (*see pp.68–69*) and WLC models to the Allied military during World War II, but the company also produced a small number of BMW-style machines for the war effort. The XA used a transversely mounted side-valve flat-twin cylinder engine, had shaft drive to the rear wheel, a four-speed gearbox, and plunger rear suspension. While it was built to the special specification of the US military, the arrival of the immensely successful four-wheel drive Willys Jeep changed the agenda as far as military motorcycle use was concerned, and only 1,000 XA bikes were ever made.

1942 XA
Although the WLA was a good all-around military bike, the US army asked Harley-Davidson to produce a shaft-drive machine and Harley turned to the enemy for inspiration. BMW's R75 was virtually cloned and a test batch of 1,000 XA's were produced, as were a batch of prototype shaft-drive Indians, the idea being that the better of the two bikes would be awarded a contract. Neither satisfied the army, which ordered extra WLAs instead, thus ending the XA's brief life.

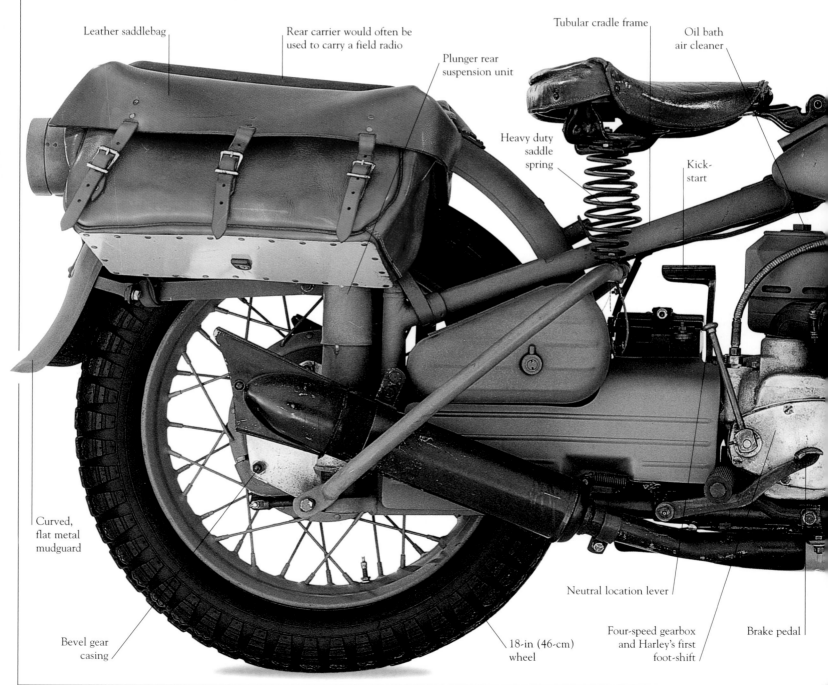

Leather saddlebag

Rear carrier would often be used to carry a field radio

Plunger rear suspension unit

Tubular cradle frame

Oil bath air cleaner

Heavy duty saddle spring

Kick-start

Curved, flat metal mudguard

Bevel gear casing

18-in (46-cm) wheel

Neutral location lever

Four-speed gearbox and Harley's first foot-shift

Brake pedal

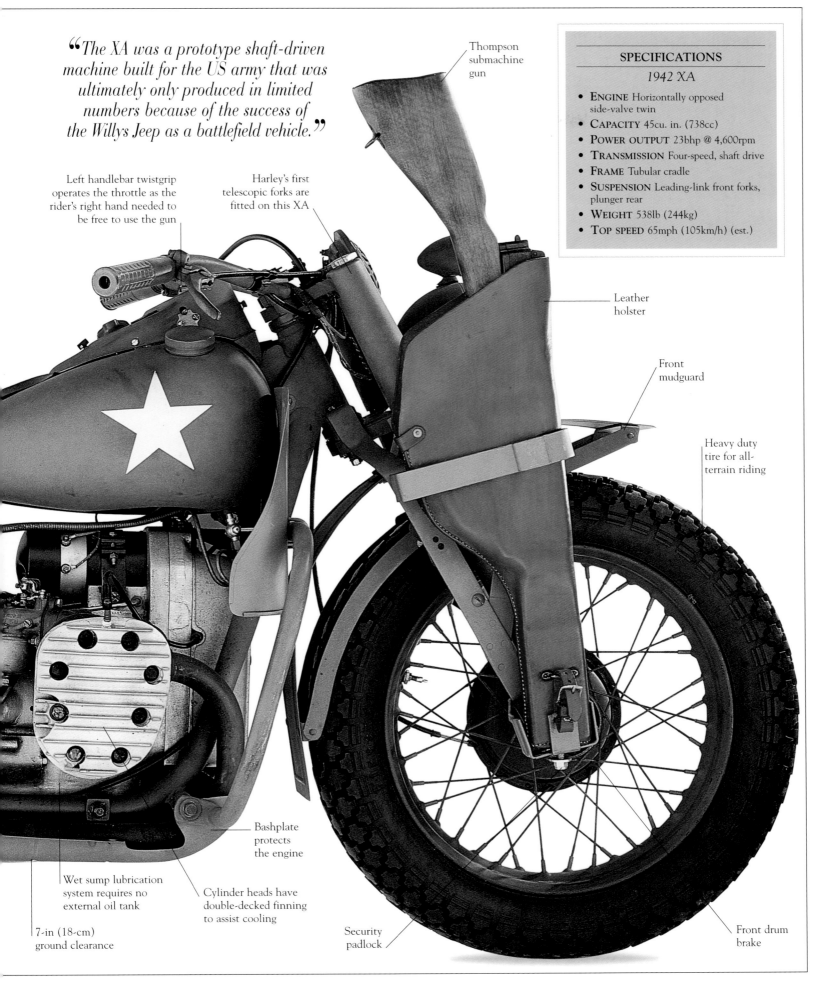

"The XA was a prototype shaft-driven machine built for the US army that was ultimately only produced in limited numbers because of the success of the Willys Jeep as a battlefield vehicle."

Thompson submachine gun

Left handlebar twistgrip operates the throttle as the rider's right hand needed to be free to use the gun

Harley's first telescopic forks are fitted on this XA

Leather holster

Front mudguard

Heavy duty tire for all-terrain riding

Bashplate protects the engine

Wet sump lubrication system requires no external oil tank

Cylinder heads have double-decked finning to assist cooling

7-in (18-cm) ground clearance

Security padlock

Front drum brake

SPECIFICATIONS
1942 XA

- **ENGINE** Horizontally opposed side-valve twin
- **CAPACITY** 45cu. in. (738cc)
- **POWER OUTPUT** 23bhp @ 4,600rpm
- **TRANSMISSION** Four-speed, shaft drive
- **FRAME** Tubular cradle
- **SUSPENSION** Leading-link front forks, plunger rear
- **WEIGHT** 538lb (244kg)
- **TOP SPEED** 65mph (105km/h) (est.)

CHAPTER THREE
SIDE-VALVES
1929–1969

1944 U NAVY

THE SIDE-VALVE ENGINE holds a special place
in American automotive history. Popular with
Henry Ford on his Model T and with rival
manufacturers Indian before Harley began using
it in 1926, it was cheap to make, rugged, and
reliable. Limited peformance meant it was soon
superseded by overhead-valve units, but Harley
continued to use the trusty side-valves long after
other manufacturers had given up on it.

A MILITARY MACHINE
*Side-valve units were used by many Allied forces
during World War II, their easy maintenance making
them ideal for battlefield conditions.*

1933 VLE

HARLEY INTRODUCED THE V-series in 1930, 14 years after rivals Indian had made their first side-valve big twins, but the bike suffered a number of teething problems. The first two months' production had to be recalled so that new frames, flywheels, engine cases, valves, springs, and kick-start mechanisms could be changed. The work was carried out for owners free of charge, but it was a costly exercise that did little for the reputation of the V-series. After the shaky start, the side-valves evolved into rugged, fast, dependable bikes, and a VLE even went on to establish the American production bike speed record in 1933 of 104mph (167km/h).

1933 VLE
The VLE was the high-compression model in the series, with magnesium-alloy pistons providing the extra power. While only 3,700 bikes were built by Harley in 1933, this still accounted for 60 percent of all motorcycles sold in the US that year.

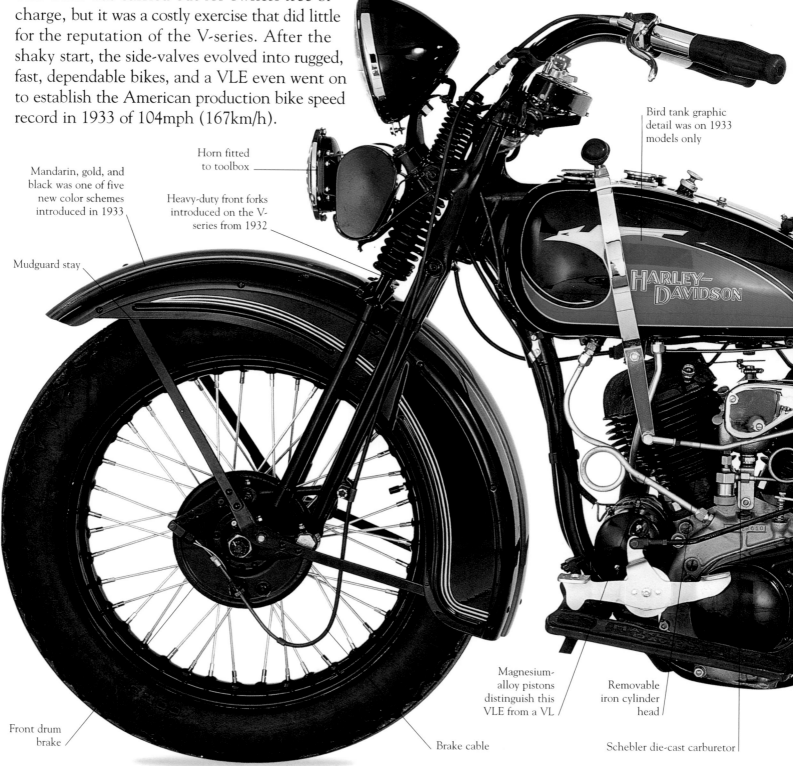

Bird tank graphic detail was on 1933 models only

Horn fitted to toolbox

Mandarin, gold, and black was one of five new color schemes introduced in 1933

Heavy-duty front forks introduced on the V-series from 1932

Mudguard stay

HARLEY-DAVIDSON

Magnesium-alloy pistons distinguish this VLE from a VL

Removable iron cylinder head

Front drum brake

Brake cable

Schebler die-cast carburetor

1930 VL

Introduced in August 1929 for the 1930 model year, this is an example of one of the all-new bikes brought in to replace the F-head V-twins. New features included the duplex primary chain, the steering head lock, and the I-beam forged fork legs. Twin headlights and the klaxon horn were carried over from the 1929 model. The bike's color scheme is the traditional olive green with vermilion striping edged in maroon and centered in gold. Drop center wheel rims were another feature introduced with the VL.

Clutch on the VL

SPECIFICATIONS
1933 VLE

- **ENGINE** Side-valve, V-twin
- **CAPACITY** 74cu. in. (1213cc)
- **POWER OUTPUT** 22bhp
- **TRANSMISSION** Three-speed (optional reverse), hand shift
- **FRAME** Tubular cradle
- **SUSPENSION** Leading-link front forks, rigid rear
- **WEIGHT** 390lb (177kg)
- **TOP SPEED** 65mph (105km/h)

> *"The V-series were reliable side-valve V-twins that consistently out-sold every other range of Harley-Davidsons during the early 1930s."*

Leather saddle with new seat-post springs for the V-series; a "buddy seat" as well as other pillion accessories could be ordered as optional extras

Chrome air filter was fitted from 1932

Though parts such as the chainguard came in black, the whole bike could be fully chromed for an extra $15

Taillight

Battery box

Rear stand

Enclosed primary drive case with automatic oiling prolonged chain life and kept the rider's boots clean

"Drop-center" wheel rims allowed modern beaded tires to be fitted

19x4-in (48x10-cm) "balloon" tires were an optional extra

1935 RL

WHEN THE ORIGINAL **D**-SERIES Harley 45s were introduced in 1929, they were nicknamed the "three-cylinder Harleys" because their vertically mounted generators resembled an extra cylinder. Harley-Davidson produced these bikes in response to the success of the popular Indian Scout, but the D-series was not considered a success and it was replaced, in 1932, by the R-series. The critical change was the new frame, which now featured a curved front downtube and allowed the fitting of a conventional horizontal generator in front of the engine. These 45s were available in four versions: the basic R model, the high-compression RL, the RLD, and sidecar RS. The Rs were replaced by the W-series in 1936. While the European market considered a 750cc machine to be a big bike, the R-series were the smallest bikes in Harley's range in the mid-1930s.

SPECIFICATIONS
1935 RL

- **ENGINE** Side-valve, V-twin
- **CAPACITY** 45cu. in. (738cc)
- **POWER OUTPUT** 22bhp (approx.)
- **TRANSMISSION** Three-speed, hand shift
- **FRAME** Tubular cradle
- **SUSPENSION** Leading-link front forks
- **WEIGHT** 390lb (177kg) (approx.)
- **TOP SPEED** 65mph (105km/h)

Even with a sprung seat post, the lack of rear suspension made for a bumpy ride

Fold-up rear mudguard section allows easy rear-wheel changes

Airflow taillight was introduced for the 1935 model year

Tubular cradle frame

Battery box

2½-in (6-cm) diameter "gas-deflecting" muffler end

Quick-detach rear hub

Exhaust pipe painted in special high-temperature black paint

New constant mesh three-speed gearbox

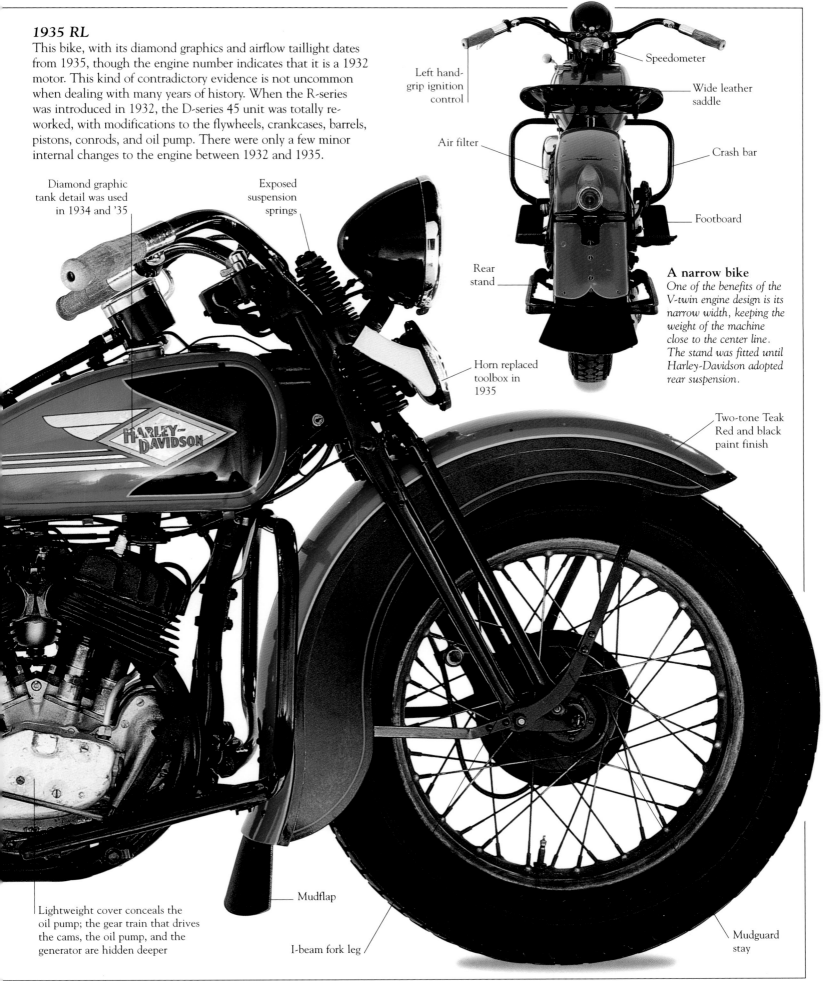

1935 RL

This bike, with its diamond graphics and airflow taillight dates from 1935, though the engine number indicates that it is a 1932 motor. This kind of contradictory evidence is not uncommon when dealing with many years of history. When the R-series was introduced in 1932, the D-series 45 unit was totally re-worked, with modifications to the flywheels, crankcases, barrels, pistons, conrods, and oil pump. There were only a few minor internal changes to the engine between 1932 and 1935.

Left hand-grip ignition control

Speedometer

Wide leather saddle

Air filter

Crash bar

Footboard

Rear stand

Diamond graphic tank detail was used in 1934 and '35

Exposed suspension springs

Horn replaced toolbox in 1935

A narrow bike
One of the benefits of the V-twin engine design is its narrow width, keeping the weight of the machine close to the center line. The stand was fitted until Harley-Davidson adopted rear suspension.

Two-tone Teak Red and black paint finish

Lightweight cover conceals the oil pump; the gear train that drives the cams, the oil pump, and the generator are hidden deeper

Mudflap

I-beam fork leg

Mudguard stay

The Flathead V-Twin

SIMPLICITY IS THE APPEAL of the side-valve, or flathead, engine. Though easy to make and maintain—there are no moving parts in the cylinder head that need to be lubricated—the downside is inefficiency. The inlet and exhaust tracts make convoluted curves so gasses have to take a long route in and out of the cylinder, resulting in poor performance and economy. However, from the 1920s to the 1950s simplicity triumphed over efficiency and the side-valve became the iconic American engine for both cars and motorcycles. Proof of its durability is found in the Servi-Car (*see pp.74–75*), which used side-valves until 1973.

Brief twin encounter

Although side-valves were used on the short-lived Sport Twin and on some singles, the 45cu. in. unit, as seen here on the RL, was its most famous application.

Sparkplug; Harley made its own rebuildable sparkplugs at the start of the side-valve era

No moving parts in the cylinder head allows the engine height to be low

Cylinder head contains specially shaped combustion chamber

Iron barrel

Valve cover conceals the valve stem, spring, and tappet adjuster

Inlet port

Inlet ports meet at the center of the V, so only a short carburetor manifold is necessary

Alloy cylinder head with vertical cooling fins; early side-valves had iron heads

Exhaust port exits downward; difficult gas flow limited the power of side-valves

Ribbed aluminum timing gear cover was introduced from 1937, designed to help cool the unit and make it more efficient

Well-used oil filler

In the side-valve era, continual engine lubrication was vital. The RL still used the constant loss system in which oil was burned or blown out of the engine without recirculating.

Filling up the oil tank and using the extra hand-operated oil pump when necessary were essential riding rituals

Outer casing contains the camshaft and timing gears

"*Rugged and reliable, these are the classic American V-twins. If you want to ride around the world, the best choice is a side-valve Harley-Davidson.*"

STEVE SLOCOMBE
(HARLEY-DAVIDSON RESTORER)

CONFIGURATION EXPLANATION

Exhaust and inlet valves are situated adjacent to the cylinder, hence the side-valve name, and because the valve stems have to be parallel, four separate cams are required. The inlet and exhaust ports are effectively L-shaped and it is these convoluted tracts that make the side-valve such an inefficient engine layout.

The oil pump is driven from the rear exhaust valve camshaft gear

THE COMPETITION

• 1930 INDIAN SCOUT •
Not so much competition as inspiration, the side-valve Indian Scout 101 was built from 1928–31 and was the benchmark 45cu. in. V-twin. It is widely regarded as the best bike Indian ever built.

Pushrod tube

Connection to oil tank

Ignition timer fits in this hole

2000s
1990s
1980s
1970s
1960s
1950s
1940s
1930s
1920s
1910s
1900s

1941 WLD Sport Solo

HARLEY-DAVIDSON HAD originally followed Indian when the latter had produced its first 45 cu. in. side-valve machine in 1927. Initially, the Indian 45s were the most highly regarded, but by the time Harley introduced its W-series in 1937, it was the Milwaukee-built bikes that enjoyed the better specification and reputation. Replacing the R-series—with which they had much in common—the three models in the original lineup were the basic W, this high-compression WLD, and the competition model WLDR. The main difference over the Rs was in the new styling, which mimicked the classy 61 Knucklehead (*see pp.78–79*) that had been introduced the previous year. Just like their big brother, the 45s now had teardrop tanks with an integrated instrument panel and curved mudguards, creating a quality range that further established Harley as the market leader.

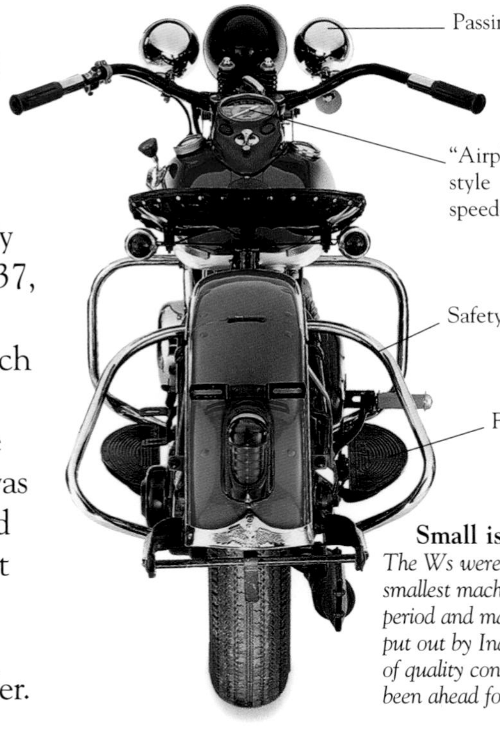

Passing lights

"Airplane"-style speedometer

Safety bar

Footboard

Small is beautiful
The Ws were Harley's smallest machines of the period and matched the 45s put out by Indian. In terms of quality control, Harley had been ahead for some time.

1941 WLD SPORT SOLO
The 1941 model shown here is rare because by this date most of Harley's production was devoted to military machines (*see WLA, pp.66–67*). The 45s were basic, robust machines that made them ideal for converting to military bikes.

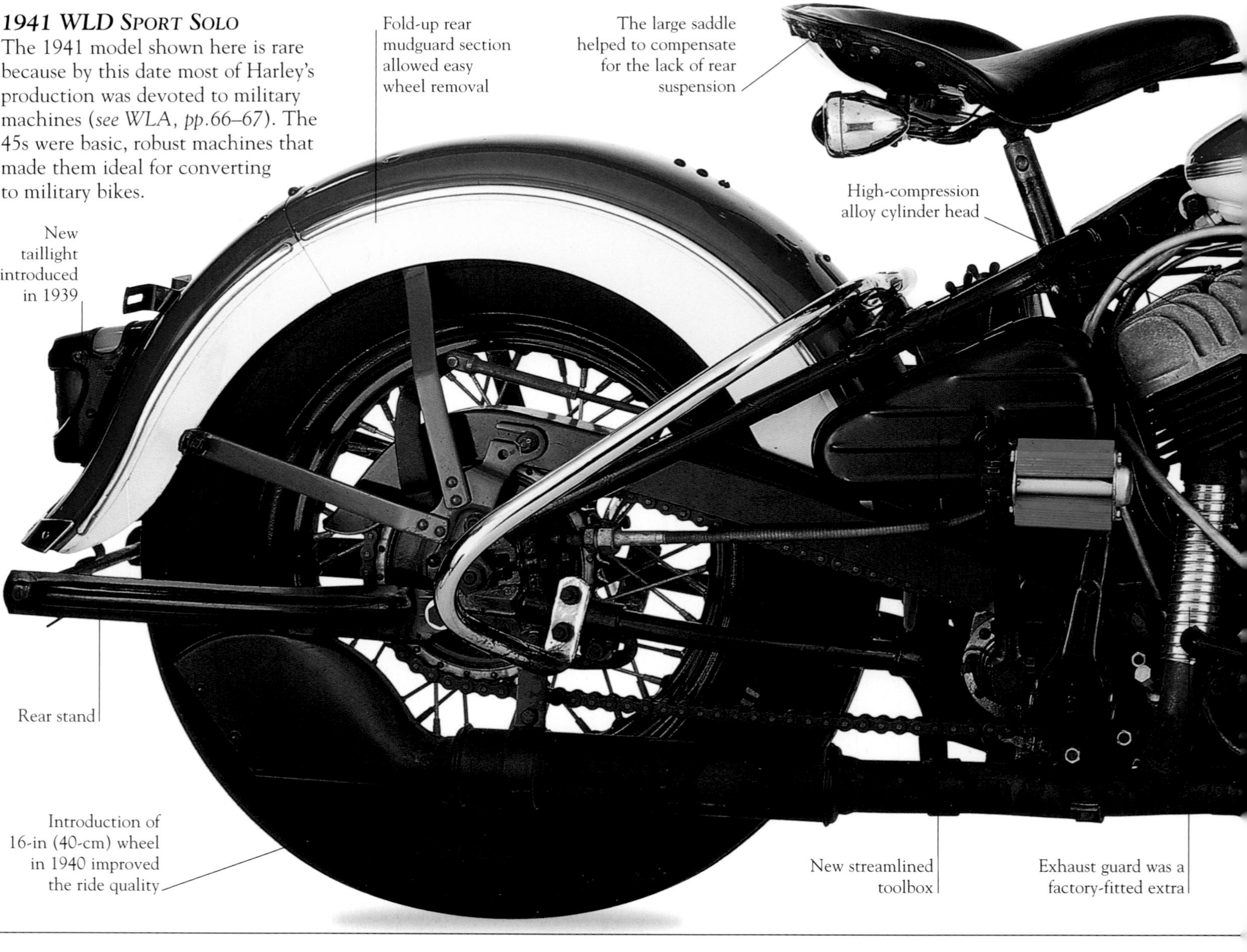

Fold-up rear mudguard section allowed easy wheel removal

The large saddle helped to compensate for the lack of rear suspension

New taillight introduced in 1939

High-compression alloy cylinder head

Rear stand

Introduction of 16-in (40-cm) wheel in 1940 improved the ride quality

New streamlined toolbox

Exhaust guard was a factory-fitted extra

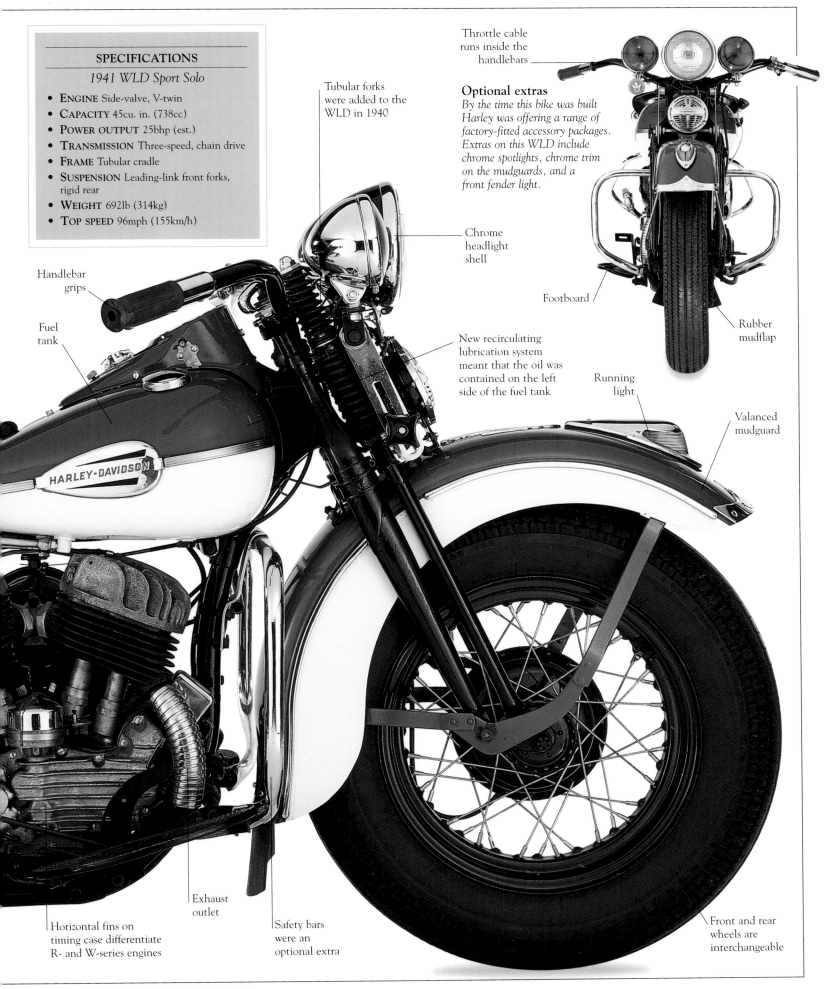

SPECIFICATIONS
1941 WLD Sport Solo

- **ENGINE** Side-valve, V-twin
- **CAPACITY** 45cu. in. (738cc)
- **POWER OUTPUT** 25bhp (est.)
- **TRANSMISSION** Three-speed, chain drive
- **FRAME** Tubular cradle
- **SUSPENSION** Leading-link front forks, rigid rear
- **WEIGHT** 692lb (314kg)
- **TOP SPEED** 96mph (155km/h)

Throttle cable runs inside the handlebars

Tubular forks were added to the WLD in 1940

Optional extras
By the time this bike was built Harley was offering a range of factory-fitted accessory packages. Extras on this WLD include chrome spotlights, chrome trim on the mudguards, and a front fender light.

Chrome headlight shell

Handlebar grips

Fuel tank

New recirculating lubrication system meant that the oil was contained on the left side of the fuel tank

Footboard

Rubber mudflap

Running light

Valanced mudguard

Exhaust outlet

Horizontal fins on timing case differentiate R- and W-series engines

Safety bars were an optional extra

Front and rear wheels are interchangeable

1942 WLA

THE MOTORCYCLE WAS A USEFUL vehicle for military dispatch and escort duty and had been used in these roles almost since its invention. Harley-Davidson had supplied a number of machines to the US military toward the end of World War I and, despite the fact that rival Indian actually contributed more bikes, the company pulled off a bit of a coup when the the first US serviceman to enter Germany in 1918 was photographed riding a Harley. The outbreak of World War II created a huge demand for two-wheeled transportation, with the military needing machines that would survive abuse and rugged terrain, and would also be simple to ride and repair. The WLA, based on the civilian WL models (*see pp.66–67*), fitted the bill perfectly. Together with the similar WLC model, manufactured for the Canadian armed forces and with only detail changes from the WLA, Harley-Davidson built over 80,000 of these models for the Allied war effort. Other Harley-Davidsons supplied to the military were the XA (*see pp.56–57*) and the U (*see pp.70–71*).

(*see pp.66–67*) (*see pp.56–57*) (*see pp.70–71*)

SPECIFICATIONS
1942 WLA

- **ENGINE** Side-valve, V-twin
- **CAPACITY** 45cu. in. (738cc)
- **POWER OUTPUT** 23bhp @ 4,600rpm
- **TRANSMISSION** Three-speed, chain drive
- **FRAME** Tubular cradle
- **SUSPENSION** Leading-link front forks, rigid rear
- **WEIGHT** 576lb (261.5kg)
- **TOP SPEED** 65mph (105km/h)

1942 WLA
The huge numbers of military Harleys made during World War II mean that the WLA is relatively common in classic bike circles. Although many are presented in their original trim, many owners have "civilianized" them by painting them in different colors and dispensing with their military accessories.

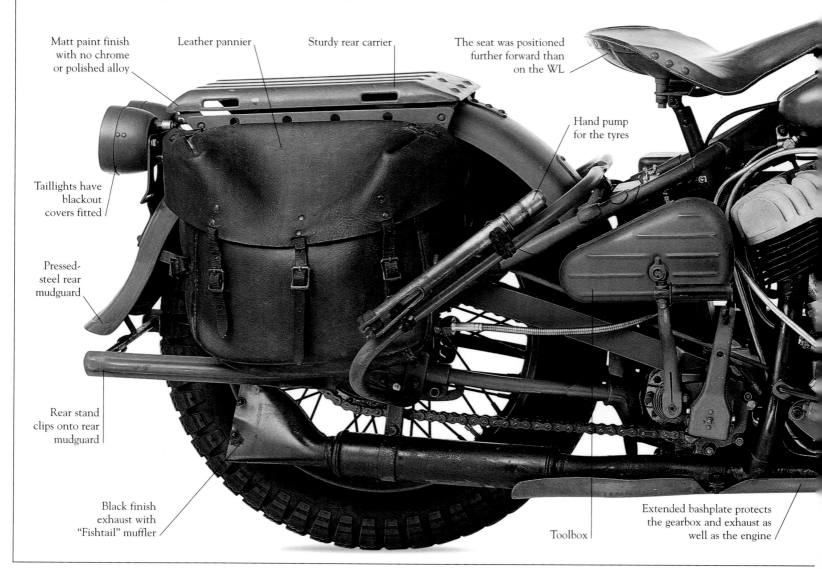

Matt paint finish with no chrome or polished alloy

Leather pannier

Sturdy rear carrier

The seat was positioned further forward than on the WL

Hand pump for the tyres

Taillights have blackout covers fitted

Pressed-steel rear mudguard

Rear stand clips onto rear mudguard

Black finish exhaust with "Fishtail" muffler

Toolbox

Extended bashplate protects the gearbox and exhaust as well as the engine

"*The rugged side-valve WLA was the most successful military motorcycle ever built, with over 80,000 seeing action for the Allied forces during World War II.***"**

Windshield fitted with canvas fairing

Rearview mirror

Instrument console

Rubber grip handlebars

Fuel filler-cap

45-caliber Thompson submachine gun

Leather machine-gun holster

Blackout lighting made night riding interesting

Small windshield

Front brake lever

Horn

Ammo box

Canvas leg-shield

Daunting front
Given this imposing view of the WLA, it's hard to believe that the military considered these bikes easy to ride, but the side-valve unit proved to be very reliable.

Mudguard is cut-down and raised to prevent clogging in muddy conditions

Synthetic material replaced rubber for tires as the war continued

Chunky block-tread tire

Front drum brake

Ground clearance was increased by extending the forks—originally on the civilian WL model—by over 2in (5cm)

ARMY

Points case

Footboard

Canvas leg-shield

1944 U Navy

THE U MODEL WAS INTRODUCED in 1937 as a replacement for the V-series 74 and 80cu. in. twins. The redesigned engine, which had a recirculating lubrication system, was fitted into a chassis taken from the 61EL Knucklehead (see pp.78–79) that had been introduced the previous year. The styling of other components including the fuel tank and running gear was also based on the sublime Knucklehead. Side-valve fans got an improved engine with a four-speed gearbox in a much more modern-looking package, and these large-capacity side-valve machines proved to be especially useful for sidecar work. With the outbreak of World War II, Harley began supplying large numbers of machines to the Allied war effort, mainly 45cu. in. bikes (see WLA, pp.68–69) but also some 74cu. in. U models. This example was supplied to the US Navy and used on shore duties on Guam.

SPECIFICATIONS

1944 U Navy

- **ENGINE** Side-valve, V-twin
- **CAPACITY** 74cu. in. (1213cc)
- **POWER OUTPUT** 22bhp
- **TRANSMISSION** Four-speed, hand shift
- **FRAME** Tubular cradle
- **SUSPENSION** Leading-link front forks
- **WEIGHT** 390lb (177kg)
- **TOP SPEED** 75mph (120km/h)

"The U model was styled on the all-new Knucklehead, but incorporated a revised version of Harley's trusty side-valve unit."

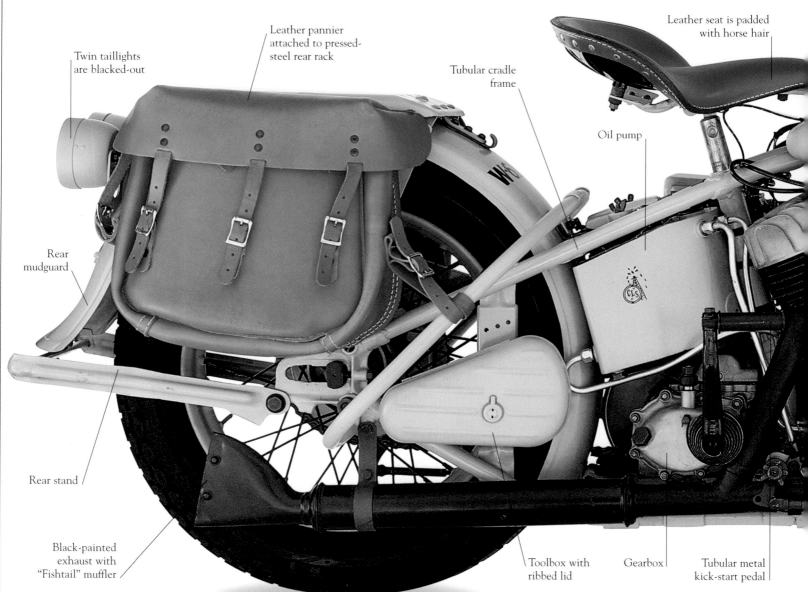

Twin taillights are blacked-out

Leather pannier attached to pressed-steel rear rack

Tubular cradle frame

Leather seat is padded with horse hair

Oil pump

Rear mudguard

Rear stand

Black-painted exhaust with "Fishtail" muffler

Toolbox with ribbed lid

Gearbox

Tubular metal kick-start pedal

1944 U Navy

The U-series bikes were simple and rugged, making them ideal for use in a military capacity. Modifications on this bike include blackout lighting and a rifle holster. War-time shortages meant that rubber handgrips were substituted by plastic and the kick-start pedal was nothing more than a bare metal tube. While the 80cu. in. version was phased out in 1941, the 74cu. in. model remained in production until 1948.

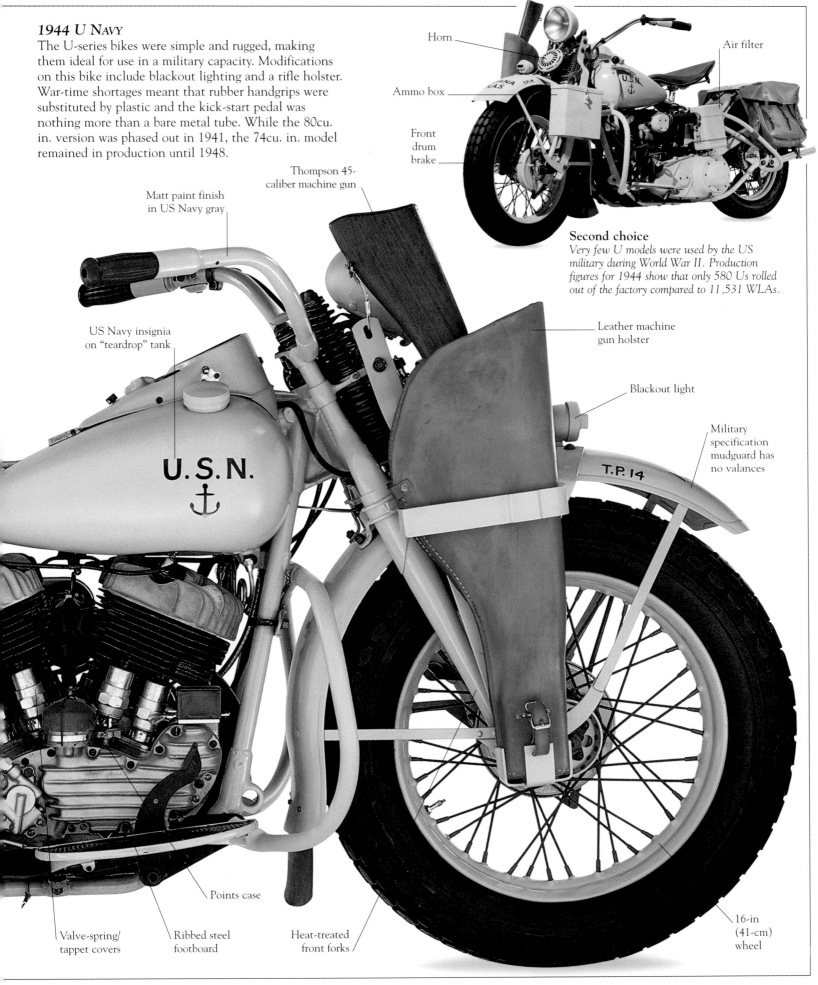

Horn

Air filter

Ammo box

Front drum brake

Matt paint finish in US Navy gray

Thompson 45-caliber machine gun

Second choice

Very few U models were used by the US military during World War II. Production figures for 1944 show that only 580 Us rolled out of the factory compared to 11,531 WLAs.

Leather machine gun holster

US Navy insignia on "teardrop" tank

Blackout light

Military specification mudguard has no valances

U.S.N.

T.P. 14

16-in (41-cm) wheel

Points case

Valve-spring/ tappet covers

Ribbed steel footboard

Heat-treated front forks

1949 WR Racer

IN 1934 THE RULES OF AMERICAN RACING were changed to encourage the participation of amateur riders on cheaper, production-based motorcycles. Though influenced by the fact that Harley-Davidson and Indian's 45cu. in. twins were comparatively inexpensive and popular at the time, the change meant that Harley had to put out some new models to meet the challenge of the class. In 1937 Harley offered the tuned WLDR, but the real response came in 1941 when the WR (flat-track) and WRTT (TT) models were introduced. These pure racing machines were supplied without any extraneous equipment—the WR, for example, came with footrests rather than boards, a lightweight frame, and no brakes. More importantly, the engine was much more powerful than the basic W models (*see pp.66–69*) on which the bike was based.

SPECIFICATIONS
1949 WR Racer
- **ENGINE** Side-valve, V-twin
- **CAPACITY** 45cu. in. (738cc)
- **POWER OUTPUT** 38bhp
- **TRANSMISSION** Three-speed, hand shift
- **FRAME** Tubular cradle
- **SUSPENSION** Leading-link front forks
- **WEIGHT** 300lb (136kg)
- **TOP SPEED** 110mph (177km/h) (est.)

This style of tank detail was introduced in 1947 and continued on the W-series through to 1951

The WR came with a choice of 18- or 19-in (46- or 48-cm) wheels

Shortened rear mudguard

Tubular cradle frame

Leather saddle pivots from the tank

Heavy-duty saddle springs

Frame bracing strut

Different sprockets allowed gearing changes

Rear drum brake was not an original component

Folding footpegs were fitted to the WR; WRTT models came with footboards, brakes, and a different frame

1949 WR RACER

The WR was available with a variety of components so riders could adapt the bike to suit their needs. Small fuel tanks could be fitted for short races, large ones for longer events; sprockets and tires also came in different sizes. Harley's real innovation with the WR was to empower amateur racers by offering them such a wealth of options, resulting in the growth in popularity of amateur bike racing.

Postwar racing victories

Harley-Davidson had phenomenal racing success in the years immediately after World War II, especially in events such as this 100-mile (161-km) road race in 1947. Harley's racing stars of the period included Babe Tancrede, winner of the Laconia 100-mile (161-km) race in 1947, and Jimmy Chann, who won the Grand National Championship in 1947, '48, and '49, as well as winning the 1949 Langhorne 100-mile (161-km) race.

Red extended control grip is a nice period touch

Exposed suspension springs

Oil filler-cap

Right side of fuel tank actually contains engine oil

Cast-iron headstock is drilled to reduce weight

Thick dirt-track racing tire

Aluminum cylinder head

Exhaust retaining spring

Two-into-one exhaust system provided optimum power

Vertical Wico magneto was derived from a unit originally intended for tractor engines

Strengthened spokes on racing wheel

Leading-link front suspension

1969 GE Servi-Car

FIRST INTRODUCED IN 1932, series G Servi-Cars were popular with US police departments through until the 1970s. They were often ridden by traffic police on parking patrol and were geared for low-speed use. The rider would drive the bike slowly past parked cars while marking the cars' tires using chalk on a stick. When the officer returned an hour later, any cars with a chalk mark on their tire would receive a ticket. Many Servi-Cars were fitted with a left-hand throttle so that the rider's right hand was free to use the chalk. They were engineered to be user-friendly as they were used by unskilled riders.

<table>
<tr><th colspan="2">SPECIFICATIONS</th></tr>
<tr><td colspan="2">1969 GE Servi-Car</td></tr>
</table>

- **ENGINE** Side-valve, V-twin
- **CAPACITY** 45cu. in. (738cc)
- **POWER OUTPUT** 22bhp
- **TRANSMISSION** Three-speed forward, one-speed reverse
- **FRAME** Tubular cradle with additional rear subframe
- **SUSPENSION** Telescopic front forks, swingarm rear
- **WEIGHT** 598lb (271kg)
- **TOP SPEED** 65mph (105km/h)

Radio antenna

Fiberglass was used for the box construction from 1967, replacing sheet metal and wood

Leather solo saddle

The gargantuan storage box was advertised as being able to carry loads of up to 500lb (227kg)

16-in (41-cm) pressed-steel wheel with chrome hubcap

Rear subframe provides necessary support for the box and extra wheel

Generator

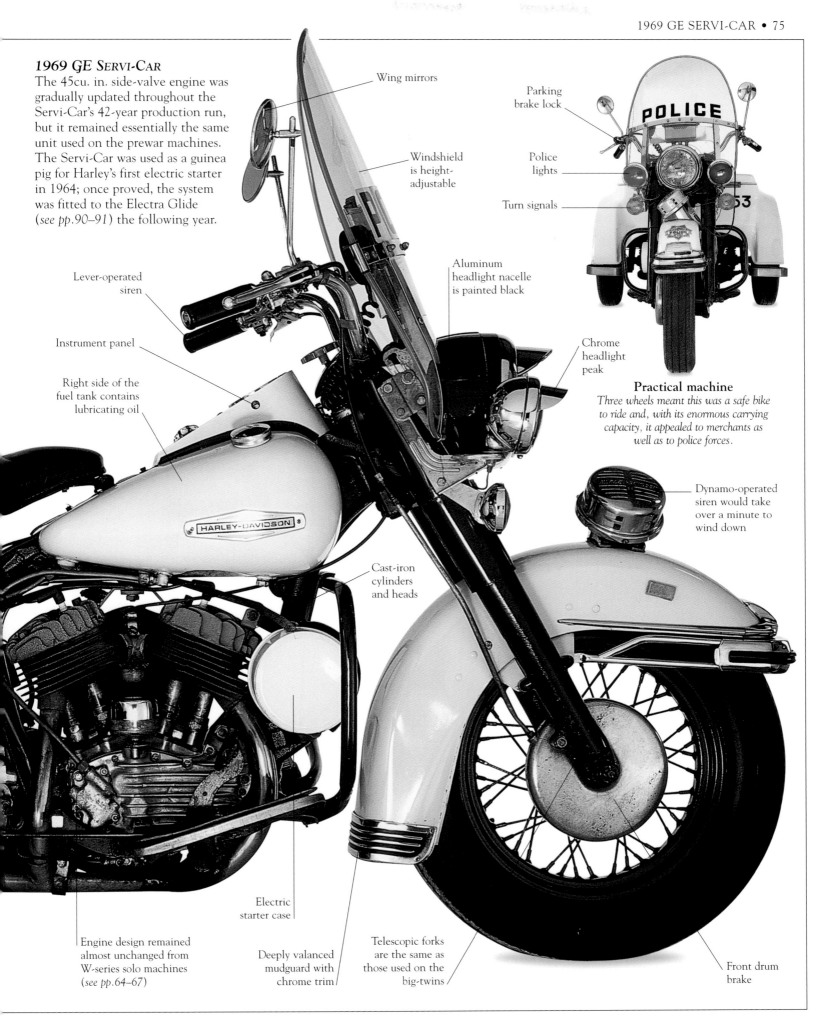

1969 GE SERVI-CAR

The 45cu. in. side-valve engine was gradually updated throughout the Servi-Car's 42-year production run, but it remained essentially the same unit used on the prewar machines. The Servi-Car was used as a guinea pig for Harley's first electric starter in 1964; once proved, the system was fitted to the Electra Glide (*see pp.90–91*) the following year.

Wing mirrors

Windshield is height-adjustable

Lever-operated siren

Instrument panel

Right side of the fuel tank contains lubricating oil

Aluminum headlight nacelle is painted black

HARLEY-DAVIDSON

Cast-iron cylinders and heads

Engine design remained almost unchanged from W-series solo machines (*see pp.64–67*)

Electric starter case

Deeply valanced mudguard with chrome trim

Telescopic forks are the same as those used on the big-twins

Front drum brake

Parking brake lock

POLICE

Police lights

Turn signals

53

Chrome headlight peak

Practical machine
Three wheels meant this was a safe bike to ride and, with its enormous carrying capacity, it appealed to merchants as well as to police forces.

Dynamo-operated siren would take over a minute to wind down

CHAPTER FOUR

OHV BIG-TWINS

1936–1984

1936 61EL "KNUCKLEHEAD"

KNUCKLEHEAD, PANHEAD, AND SHOVELHEAD can only be used as terms of endearment when discussing Harley-Davidson motorcycles. These were the nicknames given to the overhead-valve engines that powered Harley's big-twins from 1936 to 1984. Another great name in this chapter of Harley history is the Electra Glide, probably the single most famous motorbike in the world. Harley's big-twins really were the true embodiment of the American motorcycle.

STAR BIKES
Peter Fonda, seen here in the film The Wild Angels, *went on to star in the quintessential Harley-Davidson chopper film,* Easy Rider.

1936 61EL

SOME PEOPLE CONSIDER THE 61 "Knucklehead" to be the bike that put Indian out of business; others claim it was the bike that saved Harley-Davidson. Either way, this was Harley's first proper production overhead-valve twin and, introduced in 1936, it was a groundbreaking machine. The crucial new feature on the bike was its all-new overhead-valve Knucklehead engine which, for the first time on a Harley, also had a recirculating lubrication system. But the 61 wasn't just about improved technology—it was also one of the best-looking bikes that Harley ever built, and elements of its design can be seen in the cruisers of today. The teardrop fuel tank, curved mudguards, and elegant detailing gave the bike a tight, purposeful, and modern look. Although the 61EL suffered delays in development and teething troubles in production, it became one of the best-loved Harleys ever made.

SPECIFICATIONS

1936 61EL

- **ENGINE** Overhead-valve, V-twin
- **CAPACITY** 61cu. in. (1000cc)
- **POWER OUTPUT** 40bhp @ 4,800rpm
- **TRANSMISSION** Four-speed, chain drive
- **FRAME** Twin downtube tubular cradle
- **SUSPENSION** Leading-link front forks, rigid rear
- **WEIGHT** 515lb (234kg)
- **TOP SPEED** 100mph (161km/h)

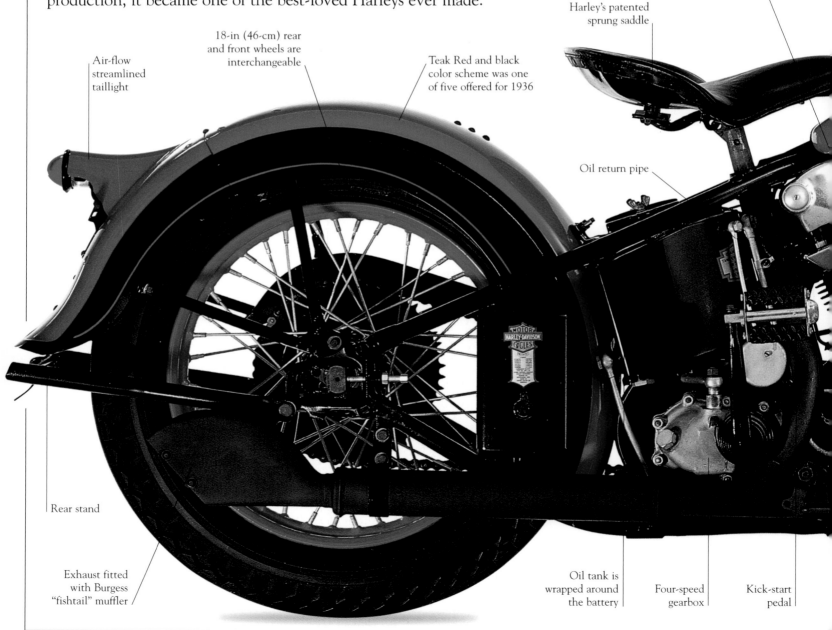

Air-flow streamlined taillight

18-in (46-cm) rear and front wheels are interchangeable

Teak Red and black color scheme was one of five offered for 1936

Harley's patented sprung saddle

Shape of the alloy rocker boxes gives the engine its Knucklehead name

Oil return pipe

Rear stand

Exhaust fitted with Burgess "fishtail" muffler

Oil tank is wrapped around the battery

Four-speed gearbox

Kick-start pedal

1936 61EL

The 61 incorporated a four-speed, constant mesh gearbox and a cradle frame. These features, combined with the new engine block, gave the bike improved performance and reliability over its side-valve predecessor. The 61 was produced in three models: 61E (medium compression), 61EL (Special Sport solo), and 61ES (medium compression sidecar). A 74cu. in. version of the Knucklehead engine (available in F, FL, and FS models) was introduced in 1941 (see pp.82–83).

Twist-grip throttle control

Large-diameter headlight

Control cables run inside the handlebars

Gear lever

Winged-face horn with chrome cover

Crash bar was a standard fitting

Beautifully sculpted fuel tank incorporates the instrument console

Front suspension springs

Teardrop fuel tank

Friction suspension damper

More innovations
The 61 carried Harley-Davidson's first four-speed, constant mesh gearbox, yet another feature that in technological terms pushed Harley ahead of its rivals Indian.

Stylish curved mudguard

HARLEY-DAVIDSON

Diagonal air intake only used on 1936 models

Points case was in this position on the big-twins until 1970

Folding footboard

Chrome molybdenum fork legs replaced the forged I-beams on the side-valve bikes

Color-co-ordinated wheel rim

The Knucklehead

THE KNUCKLEHEAD ENGINE, so named because of the shape of its rocker covers, came with two significant new features when it was introduced in 1936. The first was overhead valves, which boosted power from the V-twin unit; and the second was a single camshaft arrangement that would be used on Harley's big-twins for the next 60 years. The Knucklehead was also the first Harley-Davidson to incorporate a recirculating oil system, where oil was constantly fed through the engine rather than just burned off.

1936 61E

As well as the the 61EL, the Knucklehead was also fitted to the 61E. The difference between the models was that the EL had a high-compression Knucklehead unit, whereas the E had medium-compression.

Exhaust port

Contact breaker case

From above, the cylinders reveal a crude valve-spring enclosure and partly exposed rocker-arm shafts

Rocker shaft retaining nut; the other end is bolted to the cylinder head

Carburetor manifold

Pushrod tube

Oil-feed pipe; oil was returned to the crankcases via the pushrod tubes

Alloy cases on the top of the cylinder heads enclose the rockers

Cooling fins on cylinder barrel

Ribbed timing gear case

Engine case screw

New dry sump lubrication system meant that oil recirculated rather than just burned off

THE COMPETITION

• 1938 CROCKER •

The Crocker was produced in limited numbers in California, but it was one of the few bikes capable of outrunning a good Knucklehead. It was built in the classic style of an American V-twin and was available in the traditional red, white, and blue color scheme.

Single camshaft was superior to the twin and four-cam systems used by Harley in the past

Right there is the engine that put the Indian Motorcycle Company out of business.

LAURIE MAYERS
(KNUCKLEHEAD OWNER)

Aerodynamic fairing and fork

Disc wheel

Crankcase bolt

The Knucklehead originally had a capacity of 61cu. in., but a 74cu. in. version became available from 1941

Oil pump

WHY KNUCKLEHEAD?

Clench your fist, look at the back of your hand, and you'll see why this was called the Knucklehead. The polished nuts that retain the rocker arms are the knuckles and the pushrod tubes are the tendons running down the back of the hand. With its overhead valvegear and recirculating lubrication system, the Knucklehead was a giant leap forward for Harley-Davidson.

Aerodynamic body shell

Nationwide publicity

In March 1937, this streamlined Knucklehead achieved a record speed of 136.183mph (220km/h) when ridden by Joe Petrali at Daytona Beach, Florida. With the Knucklehead released just the year before, it was confirmation of the new engine's pedigree.

1941 74FL

THE HARLEY-DAVIDSON PHILOSOPHY of making a good idea bigger was applied to the Knucklehead engine in 1941 when capacity was increased to 74 cubic inches, though the 61cu. in. model remained in production. As well as the larger engine, the frame was stronger and adjustments had been made to the transmission since the Knucklehead first appeared in 1936 (see pp.78–81). Despite regular revamps, the evolution of Harley's big overhead-valve engine has been continuous and gradual, and there is a direct link between today's big-twins and the 1936 model. From 1942, production of the Knucklehead was badly disrupted by the war, with Harley redirecting its efforts to producing military bikes, and the FL didn't get back to being produced in significant numbers until 1946. It remained largely unchanged until the Knucklehead was replaced by the Panhead (see pp.86–87) in 1948.

SPECIFICATIONS
1941 74FL

- **ENGINE** Overhead-valve, V-twin
- **CAPACITY** 74cu. in. (1213cc)
- **POWER OUTPUT** 48bhp @ 5,000rpm
- **TRANSMISSION** Four-speed, hand shift
- **FRAME** Tubular cradle
- **SUSPENSION** Leading-link front forks
- **WEIGHT** 535lb (243kg)
- **TOP SPEED** 105mph (169km/h)

Smooth brown cowhide leather seat

Knucklehead rocker cover

Fold-up rear section allows easy wheel removal

Skyway Blue was one of five colors in the Harley range for 1941

Black painted "boat-tail" rear light

Battery is situated in the middle of the horseshoe-shaped oil tank

Rear stand

New quieter "speed-lined" muffler was introduced for 1941

Streamlined toolbox designed by Raymond Loewy

Chromed exhaust pipe covers

1941 74FL

Comparing the profile of this 1941 Knucklehead with the 1936 bike (*see pp. 78–79*), it is clear to see how much the model had developed in only five years. The streamlined toolbox, round air filter, and speed-tuned exhaust all contribute to a more modern appearance, aided by the input of designer Raymond Loewy. With design credits that included the Greyhound Scenicruiser bus and the Studebaker Avanti, Loewy was one of the most influential designers of the 20th century.

Wide handlebars

Filler-cap for 1-gallon (3.8-liter) reserve fuel tank

Imposing front
The chrome horn with embossed winged motif was introduced on the original 1936 Knucklehead and, combined with the high-mounted headlight, gave the FL a handsome front profile.

Foot-operated clutch lever

Suspension springs

"Cat's eye" instrument console

Metal tank badge introduced in 1940

Chrome horn

Wheels measuring 16 in (41 cm) were optional from 1940 and gave improved ride comfort over the larger ones

New vane-type oil pump replaced double-gear unit

7-in (18-cm) diameter circular air cleaner

Chrome mudguard trim

Whitewall tire

HARLEY-DAVIDSON

1951 74FL Hydra-Glide

HARLEY HAD BEEN KEEPING ITS riders comfortable using the springer leading-link fork (introduced 1907) and the sprung seat-post (introduced 1912) for years. By 1949, though, the merits of hydraulically damped telescopic forks were obvious and from that year they were fitted to the 61 and 74cu. in. twins. Hence the name Hydra-Glide. Another development had taken place the previous year with the arrival of the Panhead engine (*see pp.86–87*) to replace the Knucklehead. Though regarded as a classic, the Knucklehead had been prone to oil leaks and the new unit sought to address this problem—the Panhead moniker referred to the large rocker covers that enclosed the valvegear and kept the oil inside the engine. Combined with the new hydraulic valve-lifters, these modifications all amounted to reduced wear on the engine and completed another successful chapter in the history of Harley's big-twins.

SPECIFICATIONS
1951 74FL Hydra-Glide

- **ENGINE** Overhead-valve, V-twin
- **CAPACITY** 74cu. in. (1213cc)
- **POWER OUTPUT** 55bhp @ 4,800rpm
- **TRANSMISSION** Four-speed, hand shift
- **FRAME** Tubular cradle
- **SUSPENSION** Hydraulically damped telescopic forks, rigid rear
- **WEIGHT** 598lb (271kg)
- **TOP SPEED** 102mph (164km/h)

1951 74FL HYDRA-GLIDE
The L designation denotes this as a high-compression model, and the 74FL was Harley's biggest-selling model in 1951, with over 6,000 units sold. Optional foot-shift and hand-lever clutch were introduced in 1952 on the new FLF model.

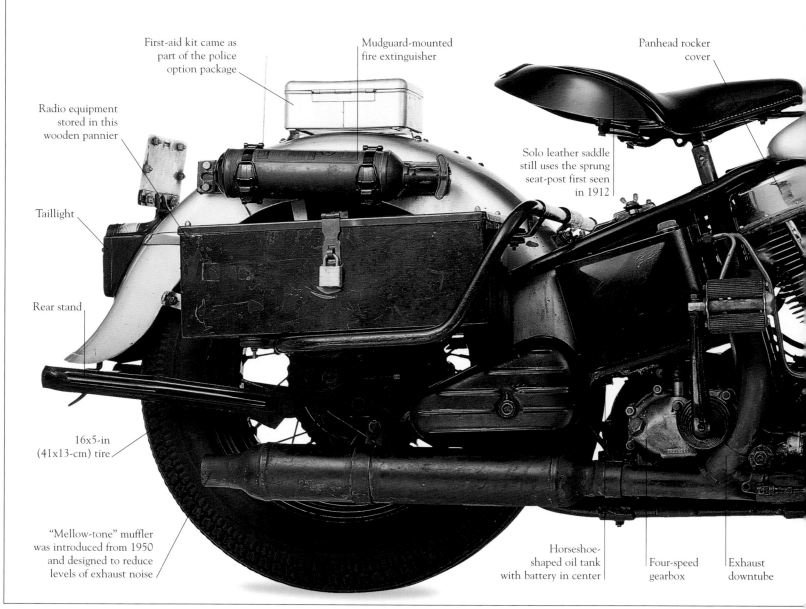

First-aid kit came as part of the police option package

Mudguard-mounted fire extinguisher

Panhead rocker cover

Radio equipment stored in this wooden pannier

Solo leather saddle still uses the sprung seat-post first seen in 1912

Taillight

Rear stand

16x5-in (41x13-cm) tire

"Mellow-tone" muffler was introduced from 1950 and designed to reduce levels of exhaust noise

Horseshoe-shaped oil tank with battery in center

Four-speed gearbox

Exhaust downtube

Windshield is
height-adjustable

Tinted lower
windshield
section

Hydraulically
damped telescopic
forks from which
the Hydra-Glide
gets its name

Large-diameter
headlight sits
in a pressed-
steel upper
fork panel

Pursuit light

Policeman's favorite
*This particular bike was used by the Willowick
Police Department, Ohio. Harley sold large numbers
of bikes to police forces and offered a standard police
accessory group costing $78.75 in 1951.*

Mudguard-mounted
police sign

Painted silver-
gray fender

Chrome
mudguard
trim

One-piece
front safety guard

Eight-rib timing
gear cover—new
for 1951

New chrome tank
badge with script
lettering and bar

Painted black
spoked wheel
and rim

8-in (20-cm)
front drum
brake

The Panhead

THERE WASN'T MUCH WRONG with the Knucklehead (*see pp.80–81*), but the Panhead is proof that you can improve on a good idea. Though the new engine was essentially a Knucklehead bottom end with a revised top section, there were a number of significant changes on the Panhead. Fully enclosed valvegear made the engine quieter and cleaner, and hydraulic valve-lifters cut down on maintenance. In addition, new aluminum cylinder heads were capped with large pan-shaped rocker covers which gave the bike its name.

Official choice
With the demise of Indian in 1953, Harleys became the only choice for police departments. Many Americans got their first close look at a motorcycle as they explained their slip-ups to a Harley-mounted highway patrol officer.

Rocker covers are lined with felt to reduce noise

Exhaust port

Points unit; timing could be altered by rotating the case

Air filter cover

Inlet manifold

New aluminum cylinder head

Pressed-steel rocker cover

Exhaust port; exhaust pipe is fitted to the heads by a clamp

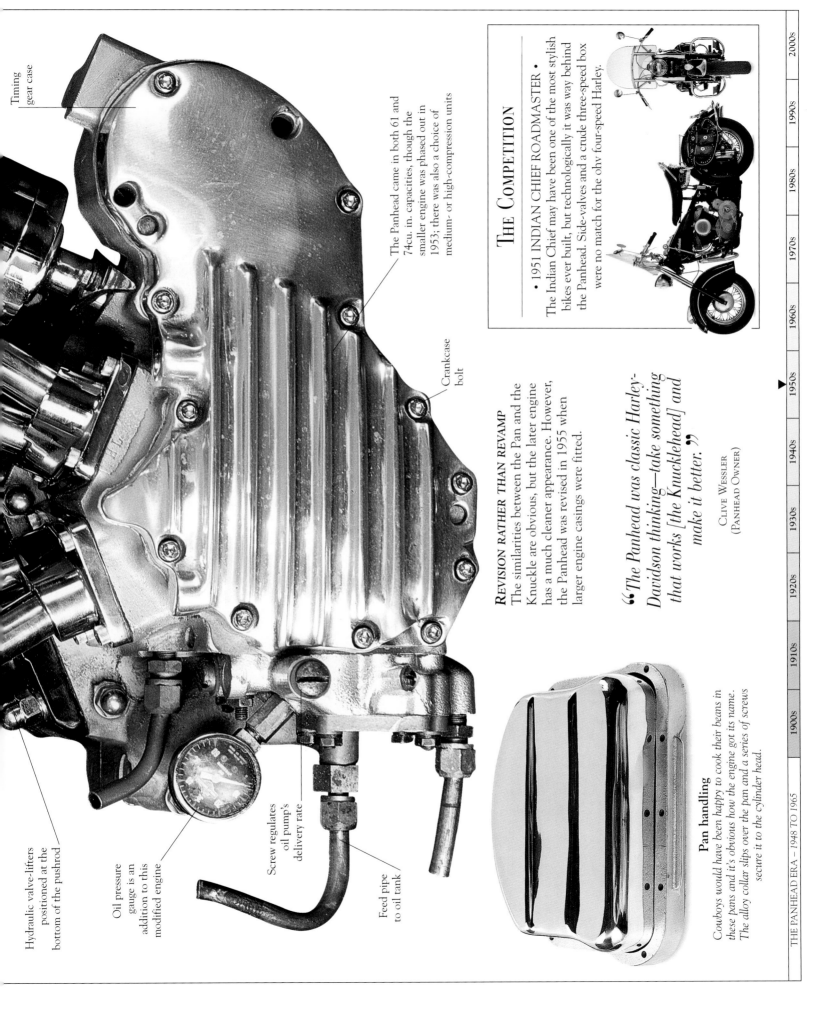

Timing gear case

Crankcase bolt

Hydraulic valve-lifters positioned at the bottom of the pushrod

Oil pressure gauge is an addition to this modified engine

Screw regulates oil pump's delivery rate

Feed pipe to oil tank

The Panhead came in both 61 and 74cu. in. capacities, though the smaller engine was phased out in 1953; there was also a choice of medium- or high-compression units

REVISION RATHER THAN REVAMP

The similarities between the Pan and the Knuckle are obvious, but the later engine has a much cleaner appearance. However, the Panhead was revised in 1955 when larger engine casings were fitted.

"The Panhead was classic Harley-Davidson thinking—take something that works [the Knucklehead] and make it better."

CLIVE WESSLER
(PANHEAD OWNER)

THE COMPETITION

• 1951 INDIAN CHIEF ROADMASTER •
The Indian Chief may have been one of the most stylish bikes ever built, but technologically it was way behind the Panhead. Side-valves and a crude three-speed box were no match for the ohv four-speed Harley.

Pan handling

Cowboys would have been happy to cook their beans in these pans and it's obvious how the engine got its name. The alloy collar slips over the pan and a series of screws secure it to the cylinder head.

2000s
1990s
1980s
1970s
1960s
1950s ▶
1940s
1930s
1920s
1910s
1900s

1960 FLH Duo-Glide

BY THE LATE 1950s, HARLEY'S big-twins had captured a section of the market for big, comfortable, large-capacity tourers. Weight wasn't an issue, but comfort and dependability were. In 1958 Harley finally added rear swingarm suspension to its Panhead big-twin and celebrated the addition with the Duo-Glide model name. By the time that this model was built in 1960, almost no two Harleys were the same, as owners tended to load their machines with extra components in order to individualize the looks and improve the comfort and capabilities of their bike. Harley offered a range of accessory groups and color schemes which meant that buyers could specify which extras they wanted on their machine when they ordered it from the dealer. These accessory-laden machines came to be known as "dressers."

Rider's handbook
Harley-Davidson's handbooks were filled with maintenance tips and other helpful hints. In addition, the comprehensive network of knowledgeable Harley dealers would always provide assistance if things went wrong.

License-plate holder

Round taillight

This model is painted in the optional color scheme of Hi-Fi Red with Birch White tank panels

Safety guards

Fiberglass rear panniers were an optional extra

Chrome exhaust cover

Cast-iron rear brake is hydraulically operated

Luggage carrier

Chrome-covered shock absorber, with the rear suspension giving the Duo-Glide its name

Passenger grab-rail

Heavy-duty saddle combined with the new rear suspension made this a perfect touring bike

Kick start

Vertical toolbox

Saddle springs were retained despite the new rear suspension

Windshield redesigned this year to incorporate new fork nacelle

Rearview mirror

"Arrow-flite" tank badge

Red tinted lower windshield section

Ball-ended handlebar lever

New aluminum fork nacelle

SPECIFICATIONS
1960 FLH Duo-Glide

- **ENGINE** Overhead-valve, V-twin
- **CAPACITY** 74cu. in. (1213cc)
- **POWER OUTPUT** 55bhp @ 7,200rpm
- **TRANSMISSION** Four-speed, hand shift
- **FRAME** Tubular cradle
- **SUSPENSION** Hydraulically damped telescopic forks, rear swingarm
- **WEIGHT** 670lb (304kg)
- **TOP SPEED** 100mph (160km/h)

1960 FLH DUO-GLIDE
The FLH designation had been introduced in 1955 for the top-of-the-range models after a revamp of the "Panhead" had produced a more efficient block resulting in greater horsepower. The H indicated that the engine had polished ports and "Victory" camshafts, which Harley claimed gave a 10-percent power increase.

Aluminum telescopic forks

Front drum brake

Chrome fender bars

Trumpet horn was fitted as standard

Ribbed timing case cover

Exhaust has a chrome cover

Rear brake pedal for new hydraulic system

Elaborate mudguard detailing

Whitewall tires

1965 74FLHB Electra Glide

IN THE EARLY 1960S ELECTRIC starters were being offered by the emerging Japanese manufacturers on even their smallest machines, so Harley felt obliged to put them on its 1200cc machines. The system was developed for the three-wheeled Servi-Car (*see pp.74–75*) in 1964 and, having proved its reliability, was attached to the FLH the following year to create the Electra Glide. A larger battery and 12-volt electrical system was needed to run the starter, so the oil tank on the FLH had to be redesigned Harley continued to put the kick-start alongside the new system for a while so that riders could impress their friends with their one-kick starting technique(and when it didn't work they could press the button).

1965 74FLHB ELECTRA GLIDE
While 1965 was the first year of the famous Electra Glide, it was also the last year for the Panhead engine. This example is a hand-shift model, which Harley continued to offer as an option until 1973. The foot clutch is on the left side, the shift-lever is on the left side of the tank, and the front brake is on the left-side handlebar. This bike is still in regular use and covers thousands of miles every year.

Seat springs

Kick-start pedal was retained on early Electra Glides

Chrome passenger grab rail

Alloy rear mudguard support

Rear license plate

Tail light

"Fishtail" muffler

Two-into-one exhaust system; a two-pipe system was an option

Chrome shock absorber cover

High-output battery situated on the right side of the frame; oil tank was moved to the left

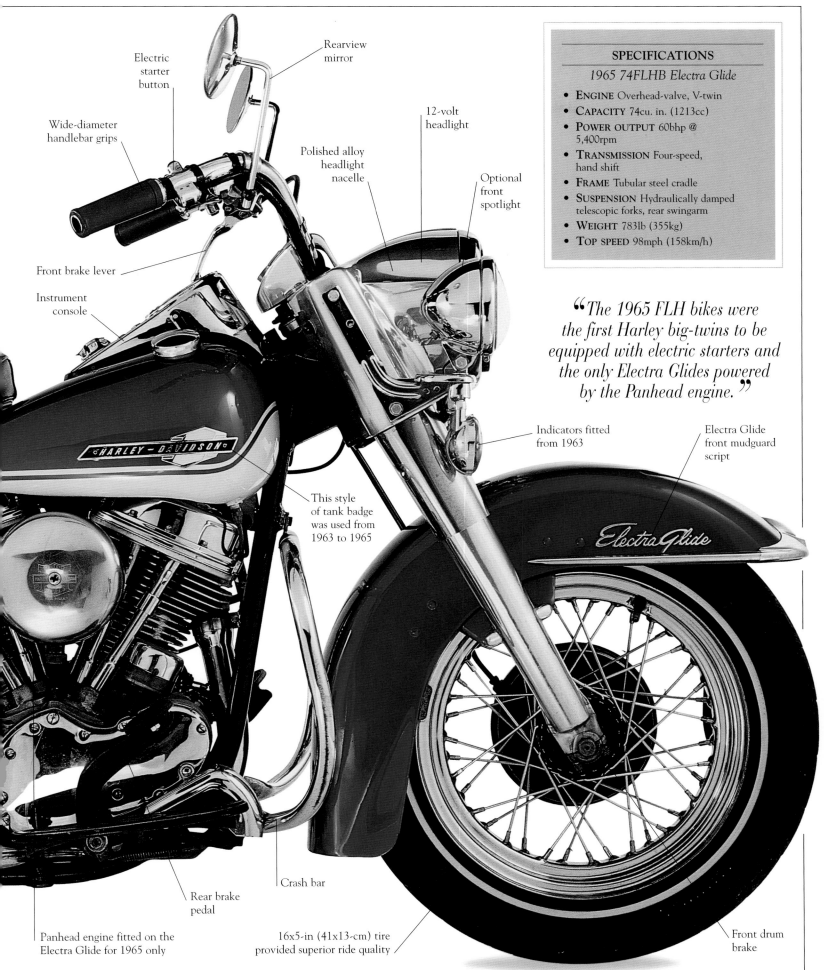

Electric starter button

Rearview mirror

Wide-diameter handlebar grips

12-volt headlight

Polished alloy headlight nacelle

Optional front spotlight

Front brake lever

Instrument console

SPECIFICATIONS
1965 74FLHB Electra Glide

- **ENGINE** Overhead-valve, V-twin
- **CAPACITY** 74cu. in. (1213cc)
- **POWER OUTPUT** 60bhp @ 5,400rpm
- **TRANSMISSION** Four-speed, hand shift
- **FRAME** Tubular steel cradle
- **SUSPENSION** Hydraulically damped telescopic forks, rear swingarm
- **WEIGHT** 783lb (355kg)
- **TOP SPEED** 98mph (158km/h)

" The 1965 FLH bikes were the first Harley big-twins to be equipped with electric starters and the only Electra Glides powered by the Panhead engine. "

Indicators fitted from 1963

Electra Glide front mudguard script

This style of tank badge was used from 1963 to 1965

Crash bar

Rear brake pedal

Panhead engine fitted on the Electra Glide for 1965 only

16x5-in (41x13-cm) tire provided superior ride quality

Front drum brake

1971 FX Super Glide

THOUGH HARLEY-DAVIDSON FROWNED on the customizers who chopped and modified its machines in the 1960s—and certainly didn't approve of the influence of the film *Easy Rider*—the company introduced its own tribute to the customizing trend in 1971. The FX Super Glide mated a kick-start 74cu. in. engine with the forks and front wheel from a Sportster to give the bike a chopper-inspired look. Styled by Willie G. Davidson, grandson of cofounder William A. Davidson, the idea was to combine the grunt of the big F-series engine with the lean looks of the X-series Sportsters. But although the Super Glide concept proved to be a winner in the long run, the unique bodywork on the 1971 model was too much for customers of the time and '72 models came with a more conventional seat and mudguard.

SPECIFICATIONS
1971 FX Super Glide

- **ENGINE** Overhead-valve, V-twin
- **CAPACITY** 74cu. in. (1213cc)
- **POWER OUTPUT** 65bhp @ 5,400rpm
- **TRANSMISSION** Four-speed, left foot-shift
- **FRAME** Tubular cradle
- **SUSPENSION** Telescopic front forks, swingarm rear
- **WEIGHT** 559lb (254kg) (with half a tank of fuel)
- **TOP SPEED** 108mph (174km/h)

3½-gallon (13.25-liter) two-part fuel tank

"Boat-tail" design was only made in 1971

Fiberglass seat unit was produced by Harley's golf-cart division

Only a small battery is needed because the FX is not fitted with an electric starter

Shovelhead rocker box

SUPER GLIDE

Muffler

The wheel rim is a non-standard alloy version

Two-into-one exhaust system

Chromed kick-start spring cover

Kick-start

Single chromed
rearview mirror

Fuel filler-cap for
the right-side of
the two-part tank

1971 FX SUPER GLIDE

This bike is an unrestored and in-use
example of the much sought-after 1971
Super Glide, with the "Sparkling America"
red, white, and blue color scheme and
"boat-tail" seat unit. These one-year-only
elements to the bike have ensured that the
'71 is now a collector's piece. This style of
seat was also an option on 1971 Sportsters.

Buckhorn,
custom-style
handlebars

Circular
taillight

Sportster-style
headlight with
"eyebrow"

Primary
drive casing

Light switch and
speedometer are
mounted in the
tank-top dash

Fold-up
footrest

Rear profile

*The slim rear profile of the Super
Glide is only compromized by the
bulbous primary drive casing.
Even so, the FX looked like no
other bike Harley had produced.*

Nonstandard
rubber fork
gaiter

16-in
(41-cm)
rear wheel

Brake
cable

Sportster
front fork

"Ham can"
air filter

Chrome cover for
master cylinder
operating the rear brake

"Cone" alternator cover
was used on the Shovelhead
engine from 1970

Single leading-shoe
front drum brake

19-in (48-cm)
front wheel

The Shovelhead

THE KNUCKLEHEAD LOOKED LIKE knuckles, the Panhead looked like pans, so the Shovelhead looked like... but you have to squint and use your imagination. Introduced in 1966, the Shovelhead's new cast-alloy rocker boxes on the revised cylinder heads gave the engine its distinctive shape. It was quieter, cleaner, more efficient, and a bit more powerful than the Panhead (see pp. 86–87). The bottom end of the engine was almost identical to that on the Panhead until a revamp in 1970 changed the electrical supply from generator to alternator

Shovelhead chopper
The Shovelhead engine is at the heart of the classic chopper. It is possible to build up a cloned Harley engine just using parts from aftermarket suppliers.

Alloy rocker cover replaces the one-piece cover used on the Panhead

Nine cooling fins indicate that this is a 74cu. in. model; 80cu. in. versions had ten fins

Rocker cover looks like a shovel turned upside down

Large "ham can" air filter

Carburetor

Rocker shafts are retained by recessed bolts

Pushrod tube

Exhaust pipes are fitted with two screws and a collar

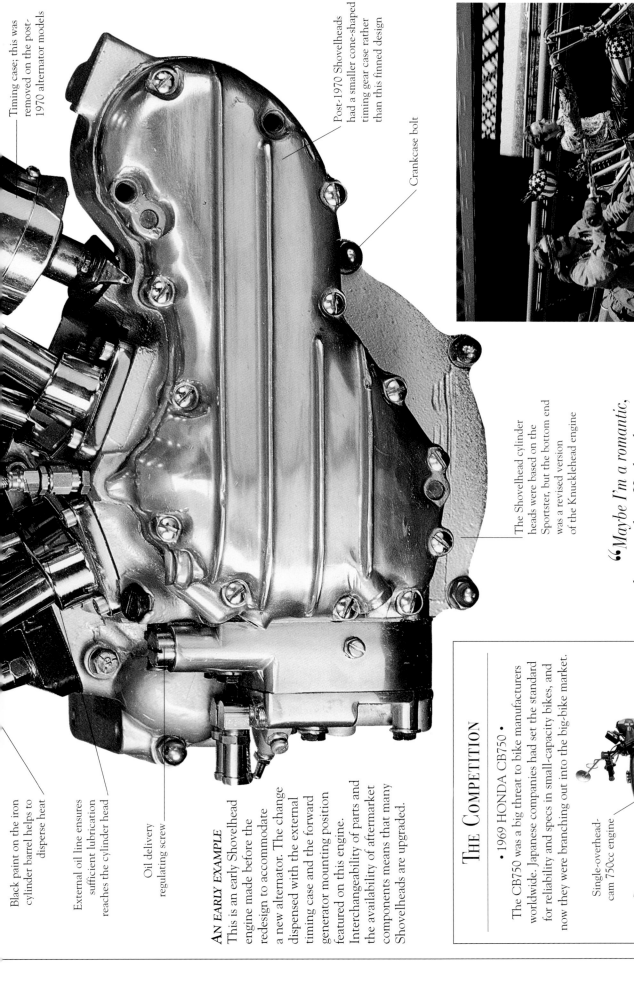

Timing case; this was removed on the post-1970 alternator models

Post-1970 Shovelheads had a smaller cone-shaped timing gear case rather than this finned design

Crankcase bolt

Black paint on the iron cylinder barrel helps to disperse heat

External oil line ensures sufficient lubrication reaches the cylinder head

Oil delivery regulating screw

The Shovelhead cylinder heads were based on the Sportster, but the bottom end was a revised version of the Knucklehead engine

AN EARLY EXAMPLE

This is an early Shovelhead engine made before the redesign to accommodate a new alternator. The change dispensed with the external timing case and the forward generator mounting position featured on this engine. Interchangeability of parts and the availability of aftermarket components means that many Shovelheads are upgraded.

"Maybe I'm a romantic, but somehow Harleys lost some charm when they replaced the Shovel. Things just got too easy."

NOEL PROBYN
(ELECTRA GLIDE OWNER)

Chopper fame

The Shovelhead era was well underway when the film Easy Rider was released in 1969. Though the bikes use Panhead (see pp.86–87) engines, the film popularized the chopper and influenced Harley's decision to build the Shovelhead FX Super Glide.

THE COMPETITION

• 1969 HONDA CB750 •

The CB750 was a big threat to bike manufacturers worldwide. Japanese companies had set the standard for reliability and specs in small-capacity bikes, and now they were branching out into the big-bike market.

Single-overhead-cam 750cc engine

1900s | 1910s | 1920s | 1930s | 1940s | 1950s | 1960s | 1970s | 1980s | 1990s | 2000s

1984 FLHX Electra Glide

AFTER **18** YEARS IN PRODUCTION the Shovelhead engine was replaced in 1984 by the new, but externally similar, Evolution engines (*see pp.144–45*). The FLHX was the swansong of the Shovelhead Electra Glides, a special limited-edition model (apparently only 1,250 were made) available in black or white with wire-spoked wheels and full touring equipment. Cynics would say that this was a good excuse to use up the last of the old-style engines, while others might argue that an engine with the reputation and life span of the Shovelhead deserved a celebratory parting shot. Either way it was the end of an era.

1984 FLHX ELECTRA GLIDE
The 80cu. in. Shovelhead engine had been introduced in 1978, and the long-established 74cu. in. version had been discontinued in 1981—three years before the 80's demise. Both had been sterling power units, but had been somewhat left behind by the new technologies being developed by foreign competition. Ultimately the FLHX was a fitting last shell for the Shovelhead block, ending an era where the old workhorse had become synonymous with the whole concept of Harley touring bikes.

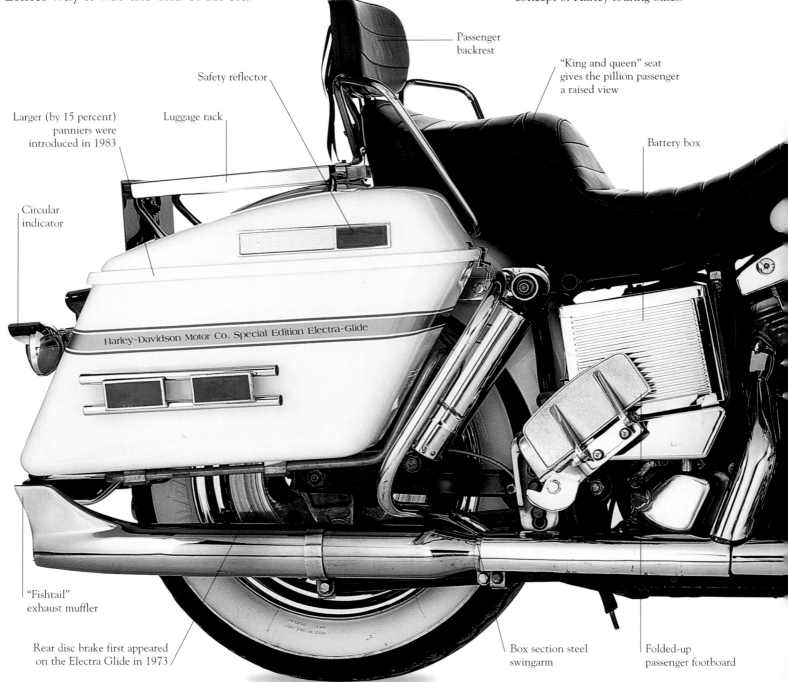

Passenger backrest

Safety reflector

"King and queen" seat gives the pillion passenger a raised view

Larger (by 15 percent) panniers were introduced in 1983

Luggage rack

Battery box

Circular indicator

Harley-Davidson Motor Co. Special Edition Electra-Glide

"Fishtail" exhaust muffler

Rear disc brake first appeared on the Electra Glide in 1973

Box section steel swingarm

Folded-up passenger footboard

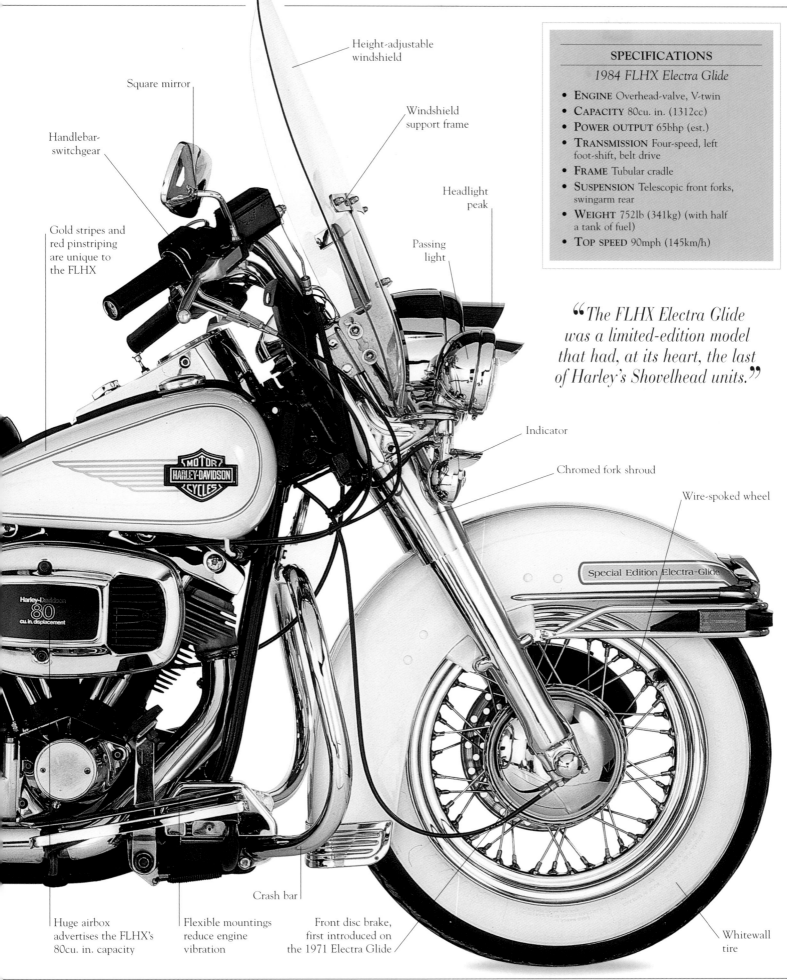

Height-adjustable windshield

Square mirror

Handlebar-switchgear

Windshield support frame

Gold stripes and red pinstriping are unique to the FLHX

Headlight peak

Passing light

"The FLHX Electra Glide was a limited-edition model that had, at its heart, the last of Harley's Shovelhead units."

Indicator

Chromed fork shroud

Wire-spoked wheel

Huge airbox advertises the FLHX's 80cu. in. capacity

Flexible mountings reduce engine vibration

Crash bar

Front disc brake, first introduced on the 1971 Electra Glide

Whitewall tire

Height-adjustable windshield

Square mirror

Handlebar-switchgear

Windshield support frame

Gold stripes and red pinstriping are unique to the FLHX

Headlight peak

Passing light

Indicator

Chromed fork shroud

Wire-spoked wheel

Huge airbox advertises the FLHX's 80cu. in. capacity

Flexible mountings reduce engine vibration

Crash bar

Front disc brake, first introduced on the 1971 Electra Glide

Whitewall tire

SPECIFICATIONS
1984 FLHX Electra Glide

- **ENGINE** Overhead-valve, V-twin
- **CAPACITY** 80cu. in. (1312cc)
- **POWER OUTPUT** 65bhp (est.)
- **TRANSMISSION** Four-speed, left foot-shift, belt drive
- **FRAME** Tubular cradle
- **SUSPENSION** Telescopic front forks, swingarm rear
- **WEIGHT** 752lb (341kg) (with half a tank of fuel)
- **TOP SPEED** 90mph (145km/h)

"The FLHX Electra Glide was a limited-edition model that had, at its heart, the last of Harley's Shovelhead units."

Special Edition Electra-Glide

Harley-Davidson 80 cu. in. displacement

CHAPTER FIVE

Post-War Small Bikes

1948–1977

1967 CRTT ALA D'ORO

LIGHTWEIGHTS, MOPEDS, AND SCOOTERS
with the famous Harley-Davidson badge?
Yes, it's true. For 30 years Harley built
lightweight machines with buzzing exhaust
notes and budget price tags, but it all ended
in failure as Harley couldn't compete with
the Japanese manufacturers. It is a fascinating
story nevertheless, involving designs acquired
as war reparations, an American scooter,
the takeover of an Italian manufacturer,
and four World Championship wins.

HARLEY-DAVIDSON CONVOY
*The Apollo 11 astronauts get a Harley-Davidson convoy through
the streets of New York in 1969. The 1960s was the decade when
Harley-Davidson produced its widest range of small motorcycles.*

1955 ST

HARLEY'S LITTLE **125**CC TWO-STROKE first appeared in the line for the 1948 season. The design was based on the German DKW, which was made available to Harley, and to the British BSA group, as part of war reparations. Harley gave the bike its own styling details, which were based on those used on the bigger models. The new bike was designated the Model S, and a 165cc version was introduced for 1953. Throughout its seven-year production run the ST retained a three-speed gearbox and rigid rear, so by the time it reached the end of its life it must have appeared very old fashioned. From 1960 Harley offered a new model based on the same engine (see *Bobcat pp.104–05*), but even these didn't get rear suspension until 1963.

> ### SPECIFICATIONS
> #### 1955 ST
> - **ENGINE** Two-stroke single
> - **CAPACITY** 165cc
> - **POWER OUTPUT** 7bhp
> - **TRANSMISSION** Three-speed
> - **FRAME** Tubular cradle
> - **SUSPENSION** Telescopic forks, rigid rear
> - **WEIGHT** 170lb (77kg)
> - **TOP SPEED** 55mph (89km/h) (est.)

1955 ST
The basic design of the ST may have been German, but Harley Americanized it by using the company's own style of motorcycle parts. The fuel tank, mudguards, seat, headlight, and other details were all derived from those used on bigger machines in the Harley line. It may not have been a big-twin, but it was a Harley-Davidson.

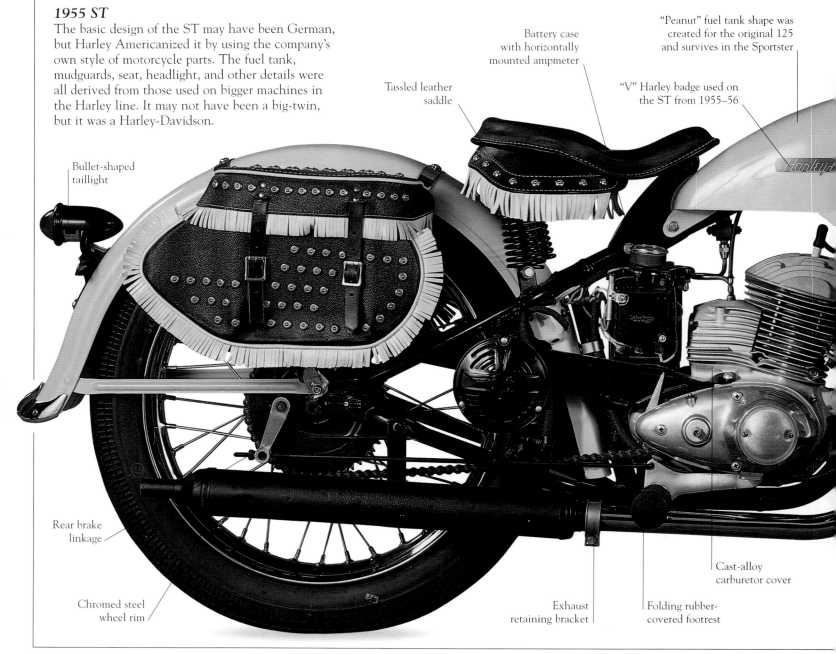

Battery case with horizontally mounted ampmeter

Tassled leather saddle

"Peanut" fuel tank shape was created for the original 125 and survives in the Sportster

"V" Harley badge used on the ST from 1955–56

Bullet-shaped taillight

Rear brake linkage

Chromed steel wheel rim

Exhaust retaining bracket

Folding rubber-covered footrest

Cast-alloy carburetor cover

"The ST was descended from the German DKW bike acquired by Harley after World War II as part of reparations."

Windshield support

Fork yoke panel incorporates the speedometer

Headlight style and mounting are derived from the Hydra-Glide (see pp.84–85)

Exhaust pipe

Valanced mudguard

Small drum brake copes with the sedate performance and light weight of the ST

German origins

DKW's influence was global in the years after World War II. Small two-strokes featuring the gear pedal and kick-start on the same axis—a DKW trademark—were produced in at least six different countries.

Lightweight "Tele-Glide" front forks first appeared in 1951

Fuel filler-cap unscrews and contains a cup for measuring oil

Optional leather panniers

Left foot-shift lever

Kick-start lever

Prop stand

Mudguard stay

Chrome front mudguard trim is nonstandard

Tinted windshield

Crash bars were optional

Light switch

Footrest

Muffler

19x3¼-in (48x8.3-cm) wheel

Extra refinements

This example is equipped with nonstandard, but period, windshield and panniers and has additional chrome trim on the mudguards.

1964 AH Topper

THE AMERICAN SCOOTER MARKET had flourished in the 1950s, but the arrival of the 165cc Topper scooter to the Harley range in 1959 heralded the end of the boom. Scooter buyers wanted bikes that were easy to ride and, though the Topper met this criterion, its boxy styling was no match for the attractive curves of the contemporary Italian Vespa and Lambretta scooters. Intriguingly, a sidecar option was offered. The Topper was dropped from the Harley catalog after the 1965 season, much to the pleasure of die-hard Harley enthusiasts.

SPECIFICATIONS
1964 AH Topper

- **ENGINE** Two-stroke, single cylinder
- **CAPACITY** 165cc
- **POWER OUTPUT** 9bhp
- **TRANSMISSION** Automatic
- **FRAME** Steel frame
- **SUSPENSION** Swingarm front and rear
- **WEIGHT** 200lb (90kg)
- **TOP SPEED** 65mph (105km/h)

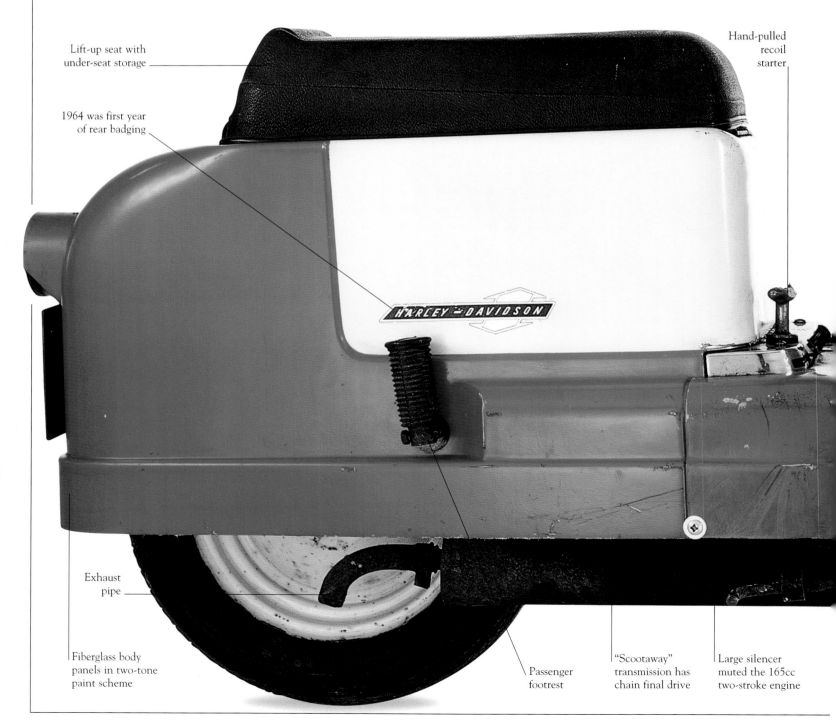

Lift-up seat with under-seat storage

Hand-pulled recoil starter

1964 was first year of rear badging

Exhaust pipe

Fiberglass body panels in two-tone paint scheme

Passenger footrest

"Scootaway" transmission has chain final drive

Large silencer muted the 165cc two-stroke engine

Front brake lever on the left handlebar had a lock on it so it could also be used as a parking brake

Handlebars were more streamlined than on previous versions

Horn grille

Apart from this Fiesta Red and white, the only other paint finish on offer in 1964 was black and white

Horn

Legshields and front mudguard were constructed of metal

Revised grips fitted for the 1962 model year

White painted headlight neacle with chrome trim

Each wheel has 5-in (13-cm) brakes

1964 AH TOPPER

While the Topper may have been a bit lacking in style, it was no worse than most other American-built machines of the era. A centrifugal clutch and belt-drive system with variable diameter pulleys provided automatic gear change and the two-stroke engine had a reed valve in the induction system. Quirkily, it was started with a pull cord, like a lawn mower.

Commuter runaround
With the Topper Harley-Davidson was trying to appeal to the urban buyer who needed a stress-free form of transportation, and it achieved this with some measure of success.

Rear brake pedal

12-in (30-cm) pressed-steel wheel

Footboard with rubber mat

Exhaust mounting bracket

Unconventional right-hand propstand

Swingarm front suspension has a shock absorber on this side only

1966 BTH Bobcat

THE AMERICAN MOTORCYCLE MARKET was shaken up in the 1960s by the arrival of vast numbers of Japanese machines. In 1963 Harley introduced the BT Pacer, a revamp of its old 165 (*see pp.100–101*) with a new frame—finally incorporating rear suspension—and a new 175cc engine. A further rework for 1966 produced the Bobcat, a one-year-only model whose styling reflected the fashion for off-road biking at the time. This was the final version of the DKW-derived two-stroke and the last Harley lightweight to be built in America—Harley's acquisition of Aermacchi in 1960 meant production shifted to Italy after 1966. Though a valiant attempt, it failed to match its well-equipped and competitively priced Japanese contemporaries.

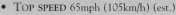

SPECIFICATIONS
1966 BTH Bobcat

- **ENGINE** Single cylinder, two-stroke
- **CAPACITY** 175cc
- **POWER OUTPUT** 10bhp (est.)
- **TRANSMISSION** Three-speed, chain drive
- **FRAME** Tubular cradle
- **SUSPENSION** Telescopic front forks, swingarm rear
- **WEIGHT** Not known
- **TOP SPEED** 65mph (105km/h) (est.)

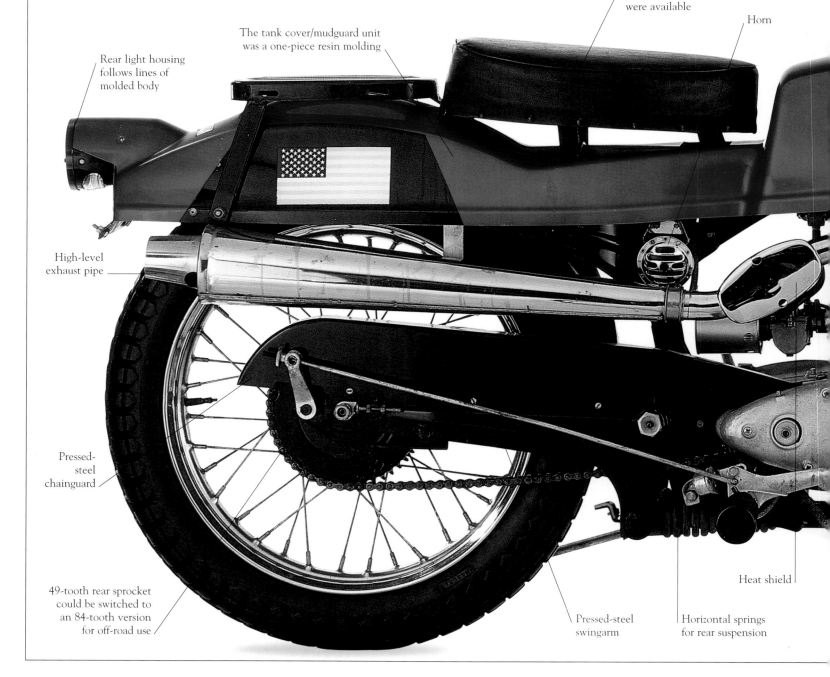

Solo and dual seat options were available

Horn

The tank cover/mudguard unit was a one-piece resin molding

Rear light housing follows lines of molded body

High-level exhaust pipe

Pressed-steel chainguard

49-tooth rear sprocket could be switched to an 84-tooth version for off-road use

Pressed-steel swingarm

Horizontal springs for rear suspension

Heat shield

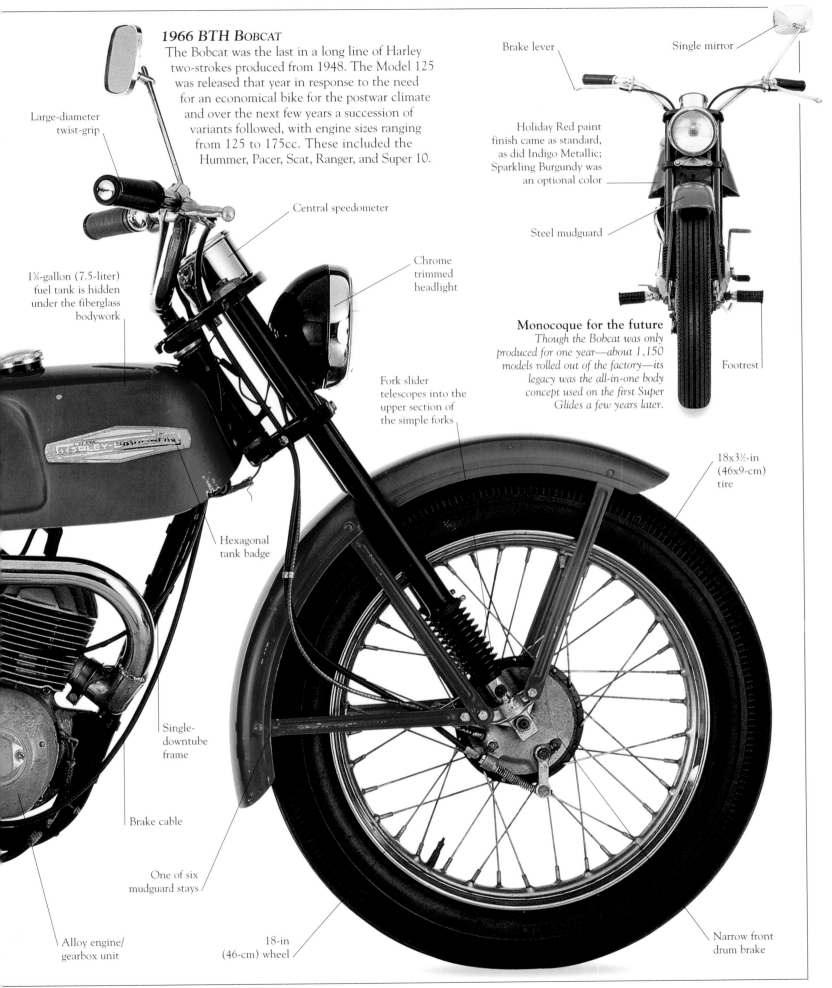

1966 BTH Bobcat

The Bobcat was the last in a long line of Harley two-strokes produced from 1948. The Model 125 was released that year in response to the need for an economical bike for the postwar climate and over the next few years a succession of variants followed, with engine sizes ranging from 125 to 175cc. These included the Hummer, Pacer, Scat, Ranger, and Super 10.

Large-diameter twist-grip

1⅞-gallon (7.5-liter) fuel tank is hidden under the fiberglass bodywork

Central speedometer

Chrome trimmed headlight

Hexagonal tank badge

Single-downtube frame

Brake cable

One of six mudguard stays

Alloy engine/gearbox unit

18-in (46-cm) wheel

Fork slider telescopes into the upper section of the simple forks

Brake lever

Single mirror

Holiday Red paint finish came as standard, as did Indigo Metallic; Sparkling Burgundy was an optional color

Steel mudguard

Footrest

Monocoque for the future
Though the Bobcat was only produced for one year—about 1,150 models rolled out of the factory—its legacy was the all-in-one body concept used on the first Super Glides a few years later.

18x3½-in (46x9-cm) tire

Narrow front drum brake

1966 Sprint H

HARLEY BOUGHT A SHARE IN THE Italian Aermacchi company in 1960 and this Harley-badged Aermacchi 250 joined the range the following year under the Sprint moniker. It was unlike any other Harley-Davidson and wary dealers treated the model with caution. The single-cylinder 246cc engine had wet sump lubrication, a cylinder positioned almost horizontally, and pushrod-operated valves. Unusually, the crankshaft rotated in the opposite direction to the wheels. Although the Sprint was a nice enough machine, it was hard-pressed to keep up with comparable Hondas of the period. Production of the Italian four-strokes continued until 1974, by which time a 350cc model had joined the 250.

SPECIFICATIONS
1966 *Sprint H*

- **ENGINE** Overhead-valve, single cylinder
- **CAPACITY** 246cc
- **POWER OUTPUT** 28bhp
- **TRANSMISSION** Four-speed, chain drive
- **FRAME** Tubular spine
- **SUSPENSION** Telescopic front forks, swingarm rear
- **WEIGHT** 280lb (127kg)
- **TOP SPEED** 90mph (145km/h) (est.)

Dual seat

Frame brace reinforces the critical area between the swingarm and the suspension top-mounting

Canister air filter

Pressed-steel mudguard is rigidly mounted to the frame

Circular taillight

License-plate mounting

Alloy wheel hub

Low-level exhaust system; earlier Sprint H models used a high-level system

Wet sump engine case

Right-hand propstand

Right-foot gearshift

1966 SPRINT H

The Sprint H was originally a trail model sold with a high-level exhaust pipe and mudguard for this popular section of the American market. The Sprint C was the roadster version. But by the time this bike was produced, a lower pipe and conventional guards were also fitted to the H. The Sprint was arguably Harley's most competent and successful small bike.

"The Sprint was one of Harley's Italian acquisitions after it bought into the Aermacchi company and was produced for 14 years."

Handlebar grip

Brake cable

Hexagonal tank badge was introduced in 1966

Steering damper

Elongated headlight shell contains the speedometer

Chrome trimmed headlight

High-level handlebars

Telescopic fork

Footrest

Front brake

Handlebar fashions
While high handlebars were popular with American buyers, European riders preferred a sportier look, so Sprints which stayed in Europe came with lower bars.

Fork gaiter

18-in (46-cm) wheel with chrome rim

2⅓-gallon (9.8-liter) fuel tank

Rocker inspection cover

Horizontal cylinder finning provided improved cooling

Dell'Orto carburetor with remote float bowl

Camshaft end cover

18x3-in (46x7.6-cm) front tire

Single leading-shoe front brake

1966 M-50 Sport

MOPEDS ALWAYS SOLD WELL IN EUROPE, but were less well suited to America, with its expanse of wide-open spaces and culture of long-distance motorcycle touring. Despite this, Harley's Italian connection resulted in the arrival of a 49cc two-stroke moped with a step-through frame and a three-speed gearbox for the 1965 season. The addition of a conventional motorcycle fuel tank and a stylish seat justified the "Sport" moniker the following year. However, poor sales saw Harley drop the M-50 after 1968 and the company never attempted mopeds again. Which was a relief to many Harley traditionalists.

1966 M-50 SPORT
Apparently Aermacchi built 10,500 M-50 Sports in 1966, though not many were brought to America. Even with a list price of $225, the ones that did get to the US didn't sell quickly. Another 15cc was added in 1967 to create the M-65 Sport, resulting in a claimed 62 percent power increase. The advertising copy stated that these bikes were "fun for young America at any age," but the reality was that America was distinctly unimpressed.

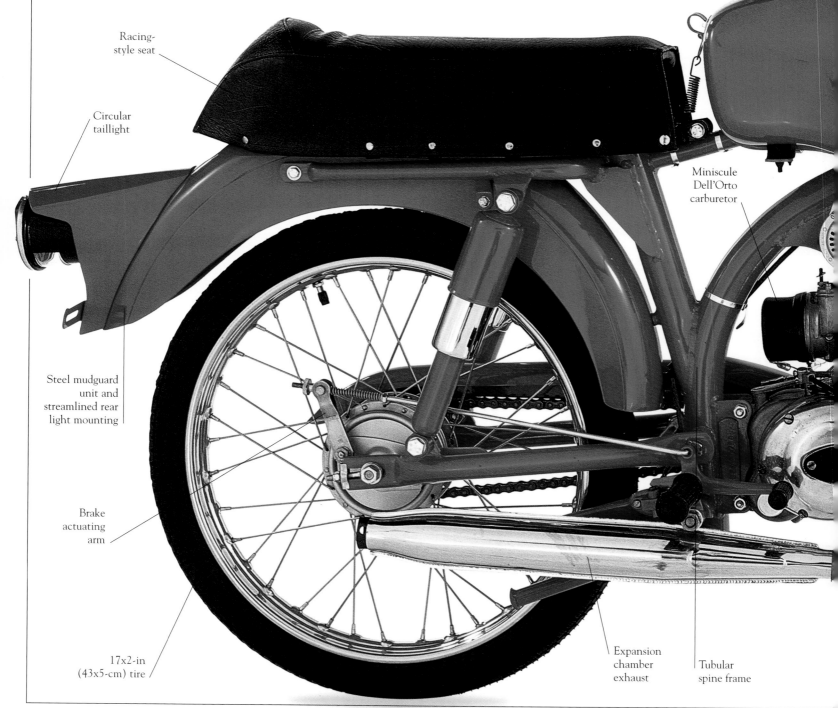

Racing-style seat

Circular taillight

Steel mudguard unit and streamlined rear light mounting

Brake actuating arm

17x2-in (43x5-cm) tire

Miniscule Dell'Orto carburetor

Expansion chamber exhaust

Tubular spine frame

Right-hand throttle with large-diameter grip

Front brake cable

Pressed-steel, elongated headlight shell with chrome trim

Rubber-mounted 2½-gallon (9.5-liter) fuel tank

HARLEY-DAVIDSON

SPECIFICATIONS

1966 M-50 Sport

- **ENGINE** Single cylinder, two-stroke
- **CAPACITY** 49cc
- **POWER OUTPUT** Unknown
- **TRANSMISSION** Three-speed with twist-grip change
- **FRAME** Tubular spine
- **SUSPENSION** Telescopic front forks, swingarm rear
- **WEIGHT** 280lb (127kg)
- **TOP SPEED** 30mph (48km/h)

❝*The M-50 was the smallest-capacity bike ever to carry the Harley-Davidson badge, but was really more at home in Italy than the United States.*❞

Shroud surround for telescopic fork

Pressed-steel front mudguard

Horn

Sparkplug cap

Ignition coil

Gear-change cable connects the left twist-grip to the change mechanism

Mudguard stay

Three-speed gearbox

Alloy cylinder head has an integral engine mounting point from which the engine is connected to the spine frame

Minimalist brakes match minimalist performance

Chromed wheel rims

1967 CRTT

THE BASIC DESIGN OF THIS Italian-built overhead-valve single was penned by Alfredo Bianchi and was originally based on a 175cc unit that powered Aermacchi's distinctive Chimera road bike. Also known as the Ala D'Oro ("Golden Wing"), a number of racing versions of the bike were produced from 1961, in 250, 350, and 402cc formats. Although the layout was the same, the race bikes differed from the road-going bikes in many respects. Engine cases were sand-cast and incorporated the provision for a dry clutch and crankshaft-driven magneto ignition.

1967 CRTT
The CRTT was only produced for one year, with just 35 rolling out of the factory. By 1967, the Italian company was pleased with the way things were going with its American partner. In 1964, for example, 75-percent of its production was being exported to the US for distribution by Harley and the range of bikes being produced was expanding.

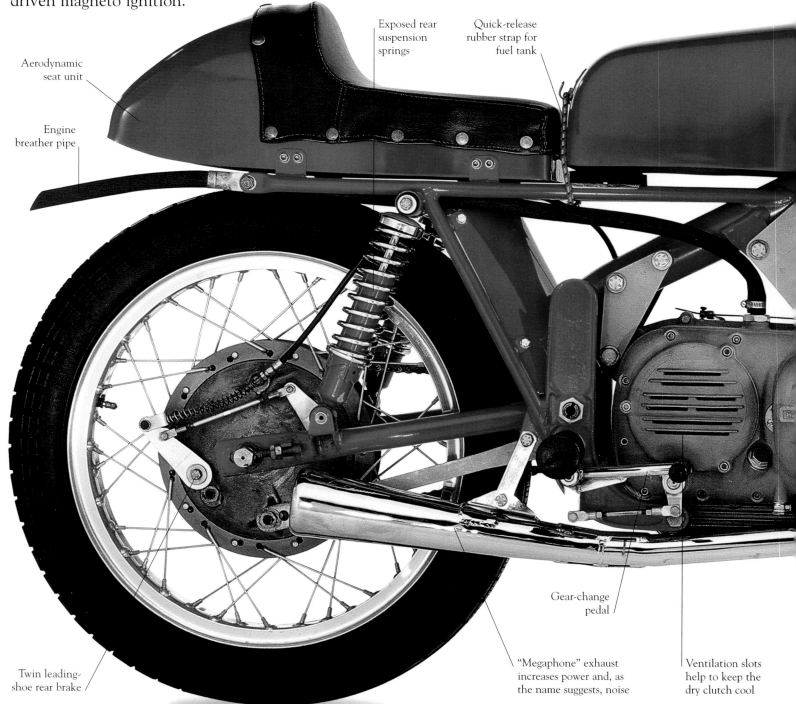

Aerodynamic seat unit

Engine breather pipe

Exposed rear suspension springs

Quick-release rubber strap for fuel tank

Twin leading-shoe rear brake

Gear-change pedal

"Megaphone" exhaust increases power and, as the name suggests, noise

Ventilation slots help to keep the dry clutch cool

"Though never an international Grand Prix winner, the CRTT was a successful national level racer—at least until eclipsed by more powerful Japanese two-strokes."

Rev counter

Fuel filler-cap

Steering damper

Elongated fiberglass fuel tank

HARLEY-DAVIDSON

Rev-counter cable

Brake cable splitting box

Fork gaiter

Italian Ceriani telescopic forks and front brakes were the best available components of the period

Brake cable

Dell'Orto carburetor

Short-stroke engine was an option

Horizontal engine cooling fins

Magneto cover and other engine casings are sand-cast magnesium alloy on these Italian-built race bikes

Flanged alloy wheel rim

Twin leading-shoe front drum brake

SPECIFICATIONS
1967 CRTT

- **ENGINE** Air-cooled, overhead-valve, horizontal single
- **CAPACITY** 248cc
- **POWER OUTPUT** 35bhp @ 10,000rpm
- **TRANSMISSION** Five-speed, chain drive
- **FRAME** Tubular spine
- **SUSPENSION** Telescopic front forks, swingarm rear
- **WEIGHT** 245lb (111kg)
- **TOP SPEED** 115mph (185km/h)

1975 250SS

IN THE MID-1970s HARLEY RELEASED a range of modern-looking single-cylinder two-strokes built in Italy at the Aermacchi factory. These replaced its ageing line-up which included the four-stroke Sprint (see pp.106–07). Offered in both street (SS) and trail (SX) styles, the bikes came in 125 and 175cc variants from 1974, and a 250cc model from 1975. The 250 may have looked a neat bike, but once again it couldn't match the strong Japanese competition of the time. The SX had some success and was produced until 1978, but only 1,417 SSs were sold in 1976 and this model was dropped soon after its release. Along with the whole Aermacchi subsidiary, which Harley decided to relinquish in 1978.

1975 250SS

The 250SS and its smaller siblings had brief production runs as Harley-Davidsons, but continued under another name after Harley sold Aermacchi in 1978. The factory and production rights were bought by Cagiva, which went on to acquire the famous Ducati and MV Agusta marques. This was not before the company achieved considerable success with the lightweight two-strokes that it acquired from Harley-Davidson.

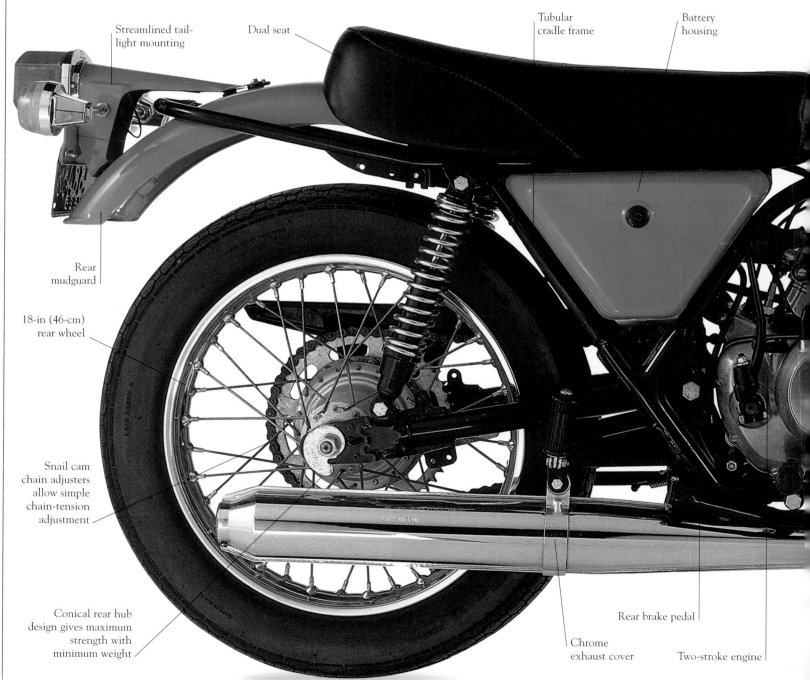

Streamlined tail-light mounting

Dual seat

Tubular cradle frame

Battery housing

Rear mudguard

18-in (46-cm) rear wheel

Snail cam chain adjusters allow simple chain-tension adjustment

Conical rear hub design gives maximum strength with minimum weight

Rear brake pedal

Chrome exhaust cover

Two-stroke engine

Rearview mirror

Clutch lever

Separate oil tank is in the top tube of the frame

Speedometer and tachometer

Styling contained a hint of traditional Harley, but the result was mainly a clean 1970s lightweight look

Fuel filler-cap for 2⅘-gallon (10.6-liter) fuel tank

Headlight switch

Betor front fork

Brake cable

Safety reflector

Exhaust retaining spring

Speedo cable

Dell'Orto carburetor

Five-speed gearbox

5⅓-in (13.5-cm) leading-shoe drum brake; later models had disc brakes

19-in (48-cm) front wheel

Chromed steel rim

SPECIFICATIONS
1975 250SS

- ENGINE Two-stroke, single-cylinder
- CAPACITY 243cc
- POWER OUTPUT Unknown
- TRANSMISSION Five-speed, chain drive
- FRAME Tubular cradle
- SUSPENSION Telescopic front forks, swingarm rear
- WEIGHT 245lb (111kg)
- TOP SPEED 85mph (137km/h)

"The 250SS was Harley's attempt to update its Sprint range but was a commercial failure due to the high quality of contemporary Japanese bikes of the same size."

1976 RR250

MOST TRADITIONAL HARLEY riders may not realize that the company won a string of World Championships in the mid-1970s, and if they do they probably don't really care. The bikes that gave Harley the titles were as far removed from the traditional V-twin as it's possible to get. These high-revving two-stroke twins were developed in Italy to take on the Japanese manufacturers in international road-racing championships. Ridden by Italian ace Walter Villa, the twins won three straight 250cc World Championships in 1975, 1976, and 1977, and a 350cc version also took that title in 1977.

SPECIFICATIONS
1976 RR250

- **ENGINE** Two-stroke twin-cylinder
- **CAPACITY** 246cc
- **POWER OUTPUT** 53bhp
- **TRANSMISSION** Six-speed, chain drive
- **FRAME** Tubular cradle
- **SUSPENSION** Telescopic front forks, swingarm rear
- **WEIGHT** 240lb (109kg)
- **TOP SPEED** 140mph (225km/h)

Aerodynamic seat hump

Molded racing seat

Knee cutout

Ignition coil

Rear shock-absorber mounting

Rear shock-absorber

Box section swingarm

Fuel tap

Twin leading-shoe rear drum brake is cable-operated

Tubular cradle frame

Eccentric swingarm mountings provide chain adjustment

Right-foot gear-change with reversed lever

Exposed dry clutch

1976 RR250

The RR250 was a production version of the bike that won four World Championships for Villa. The two-stroke twin dominated racing in the 125, 250, and 350cc classes at that time, though in most cases the bikes had a Yamaha sticker. When Kawasaki developed its disc-valve racers in the late 1970s, however, the RR250 was soon outclassed.

Bare essentials

Racing machines have minimal instrumentation, allowing the rider to concentrate on the job in hand, and the RR250 is no exception. The large, white-faced rev-counter gives all the information needed to time gear-changes to perfection.

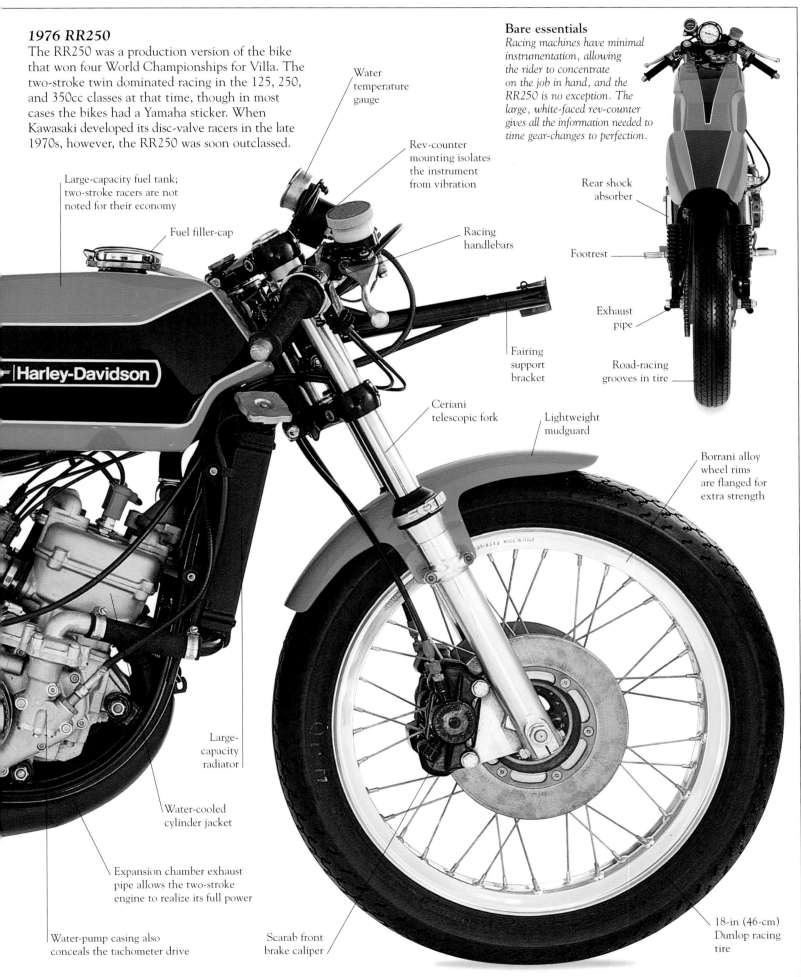

Water temperature gauge

Rev-counter mounting isolates the instrument from vibration

Large-capacity fuel tank; two-stroke racers are not noted for their economy

Fuel filler-cap

Racing handlebars

Harley-Davidson

Rear shock absorber

Footrest

Exhaust pipe

Road-racing grooves in tire

Fairing support bracket

Ceriani telescopic fork

Lightweight mudguard

Borrani alloy wheel rims are flanged for extra strength

Large-capacity radiator

Water-cooled cylinder jacket

Expansion chamber exhaust pipe allows the two-stroke engine to realize its full power

Water-pump casing also conceals the tachometer drive

Scarab front brake caliper

18-in (46-cm) Dunlop racing tire

CHAPTER SIX

SPORTSTERS

1952–2010

1972 XRTT RACER

THE HARLEY-DAVIDSON SPORTSTER is the motorcycle in its purest form—just a handsome V-twin engine, a minimalist chassis, wheels, handlebars, and a fuel tank. It represents the true experience of motorcycling, a visceral embodiment of the things that make us love motorcycles. The Sportster is the longest-running production motorcycle in the world, having been around since 1957. In that time it has been uprated and improved without altering the essential ingredients that make it what it is.

RACING PEDIGREE

Harley Sportsters have been winning races ever since their introduction, with bikes ranging from the KRTT to the XR750 taking the honors on tracks all over the world.

1952 Model K

IN THE EARLY **1950s** American motorcylists wanted more from their motorcycles. They were buying faster, better handling and better looking British imports rather than Harley's traditional 45cu. in. V-twins. In 1952 Harley hit back with the K model, its first significant new machine since the Knucklehead of 1936 (*see pp.78–79*). While the bike incorporated several novel features on a 45cu. in. side-valve block, it still could not match the performance of the smaller-capacity British machines. Increasing capacity to 54cu. in. on the KH in 1954 helped, but there was only so much that could be done with the side-valve layout. The solution came in the form of the OHV XL Sportster (*see pp.120–21*).

Rubber-mounted "buckhorn" style handlebars

Single mirror

Ignition and light-switch mounting

Hydraulically damped telescopic forks with chrome shrouds

Large-diameter headlight with chrome shell

Painted steel mudguard

8-in (20-cm) diameter drum brake

Harley-Davidson

1952 MODEL K
Innovations on the K included an engine and gearbox as one unit, swingarm rear suspension on the cradle frame, and telescopic forks. The British style of hand clutch and right foot-shift replaced Harley's usual foot clutch and hand gear-change.

SPECIFICATIONS
1952 Model K

- **ENGINE** Side-valve, V-twin
- **CAPACITY** 45cu. in. (738cc)
- **POWER OUTPUT** 30bhp (est.)
- **TRANSMISSION** Four-speed, chain drive
- **FRAME** Tubular cradle
- **SUSPENSION** Telescopic front forks, swingarm rear
- **WEIGHT** 400lb (181kg)
- **TOP SPEED** 85mph (136km/h) (est.)

Brake lever now mounted on the right

Clutch operated by lever on handlebar

Panniers came as part of a Harley accessory group for the Model K

"The forerunner of Harley's Sportster series, the Model K was nevertheless an underpowered bike."

Crash bar was an optional fitting

Air horn

Classic Harley design
While the K model was not in itself a great success, its styling has been used on Harley-Davidsons ever since; note the striking visual similarities between this bike and the 1999 model 1200 Sportster (see pp.134–35).

4½-gallon (20.5-liter) fuel tank

Cast-alloy cylinder head

Battery case with access door; the oil tank is on the other side

Saddle incorporates sprung seat-post despite the bike also having swingarm rear suspension

Swingarm rear suspension

Taillight

Rear brake pedal

Passenger footrest

Alloy primary drive case

19-in (48-cm) wheel is higher and narrower than on previous models

Cast-iron rear drum brake

1957 XL

HARLEY FINALLY FITTED OVERHEAD cylinder heads to its smaller V-twin in 1957 to create the Sportster, a model that was to become one of the longest surviving production motorcycles in the world. The Sportster combined the good looks of the earlier K (*see pp.118–19*) and KH models with enough power to match the performance of contemporary imported bikes. Although capacity remained at 54cu. in., the same as for the KH, the larger bore and shorter stroke resulted in increased horsepower. It meant that American buyers could now invest in a domestic product without suffering the indignity of being blown away by their Triumph- and BSA-mounted friends. As with the bigger twins, the factory offered a variety of accessories including the windshields, racks, panniers, crash bars, and spotlights fitted to this 1957 model. Other owners went the opposite way and stripped superfluous parts off their machines to improve both looks and performance.

SPECIFICATIONS
1957 XL

- **ENGINE** Overhead-valve, V-twin
- **CAPACITY** 54cu. in. (883cc)
- **POWER OUTPUT** 32bhp @ 4,200rpm
- **TRANSMISSION** Four-speed, chain drive
- **FRAME** Tubular cradle
- **SUSPENSION** Telescopic front forks, swingarm rear
- **WEIGHT** 463lb (210kg)
- **TOP SPEED** 92mph (148km/h) (est.)

1957 XL
When the Sportster was introduced it was intended for touring riders as well as those for whom performance was important. As a result, panniers and racks were included as factory extras, to allow Sportster riders to dress up their bikes.

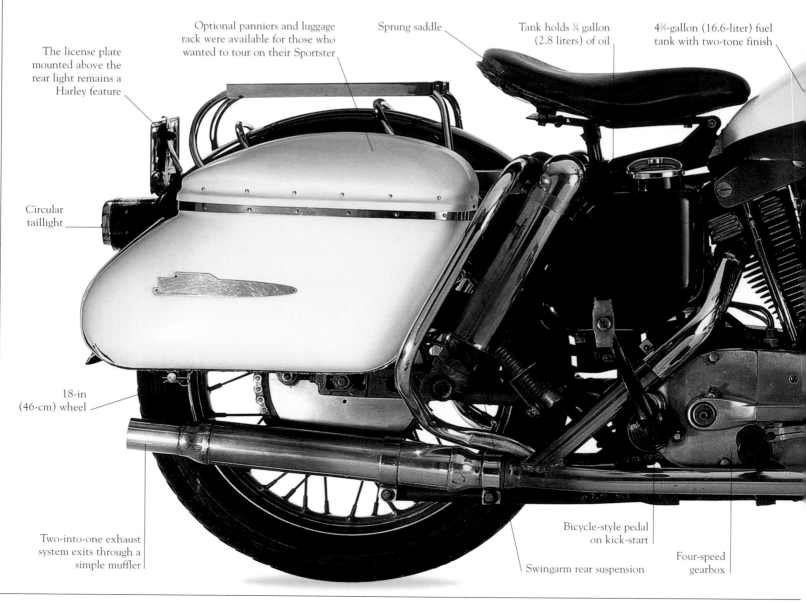

Optional panniers and luggage rack were available for those who wanted to tour on their Sportster

Sprung saddle

Tank holds ¾ gallon (2.8 liters) of oil

4⅖-gallon (16.6-liter) fuel tank with two-tone finish

The license plate mounted above the rear light remains a Harley feature

Circular taillight

18-in (46-cm) wheel

Two-into-one exhaust system exits through a simple muffler

Bicycle-style pedal on kick-start

Four-speed gearbox

Swingarm rear suspension

Rearview
mirror

Windshield was a
Harley-Davidson
optional extra

Blue-tinted lower
windshield section

The Sportster look
*The classic Sportster look developed from these
early models when an optional 2¼-gallon
(8.5-liter) "peanut" fuel tank was introduced
on XLC Sportsters in 1958. Once that was
in place, the look remained the
same up to this present day.*

❝*The XL was the first true Harley
Sportster, with overhead-valves
providing the power and revised
styling providing the looks to compete
with the British competition.*❞

Round plastic
tank emblem was
only used in 1957

Speedometer is
mounted in the
metal fork shroud

Additional
passing light

When the Sportster range
was expanded in 1958, the
XLC featured cut-down
mudguards and no lights

Optional
crash bar

Trumpet
horn

Chromed
spokes and
wheel rims

Ribbed
front tire

Two-into-one
exhaust system

Iron cylinder
head and barrel;
alloy heads were
not fitted until
the 1980s

British-style
right footshift

Single, leading-
shoe drum brake
engaged by lever
on right handlebar

1961 KRTT

ALONG WITH THE INTRODUCTION of the new K-series road bike in 1952 (*see pp.118–19*) Harley-Davidson also released a racing version, designated the KR. The bike's engine had all the tweaks you would expect in a competition power unit, while looking externally similar to the K. There were big valves, racing cams, and new bearings, as well as reshaped ports and a revised cylinder head. Riders like Brad Andres, Joe Leonard, Carroll Resweber, George Roeder, and Roger Reiman achieved many wins on the side-valve KR racers in the 1950s and 1960s. Because of the variety of track surfaces and conditions found in American racing, both sprung and rigid versions of the KR's frame were produced.

Rev-counter

Large fuel tank for long-distance races

Telescopic forks are based on those used on the road-going Sportster and K-series machines

Alloy wheel rim is lighter and stronger than the usual steel examples

Clip-on handlebar is used on long straights to reduce drag

Magneto

Tubular frame cradle

Ventilated brake drum

1961 KRTT
The basic KR was intended for dirt-track racing and so did not come equipped with brakes or suspension. This TT version had both and was ridden to victory at Daytona in 1961 by Roger Reiman at the first 200-mile (322-km) race to be held at the new banked oval track.

Wide bars for dirt-track racing

Number 55 was Roger Reiman's race number at Daytona in 1961

Side-valve success
While the standard road-going K-series was discontinued for 1957, the racing KRs had continued success with this engine layout until the late 1960s.

Air filter

Tank breather-pipe

Simple saddle is still sprung despite the rear suspension now fitted to Harley bikes

"Despite the limitations of the side-valve layout, the KR racers continued Harley's tradition of success on the race track."

Race number plate

Cut-down alloy mudguard

SPECIFICATIONS
1961 KRTT

- **ENGINE** Side-valve, V-twin
- **CAPACITY** 45cu. in. (750cc)
- **POWER OUTPUT** 50bhp
- **TRANSMISSION** Four-speed, chain drive
- **FRAME** Tubular cradle
- **SUSPENSION** Telescopic front forks, rear swingarm
- **WEIGHT** 320lb (145kg)
- **TOP SPEED** 125mph (233km/h)

Brake pedal has been drilled to reduce the bike's weight

Mudguard support

Pressed-steel primary drive cover

Tire is screwed to the rim for added security

Block tread race tire

1972 XRTT

ROAD-RACING WAS A RARITY IN America in the 1960s. Most racing action was on the dirt tracks for which the XR and KR models were conceived, but Harley decided to build a road-race version of the XR. Unlike the flat-track bike, it was fitted with a front brake and a large-diameter four-leading shoe drum, which was combined with a rear disc as on the dirt bike. Most machines had the disc-drum combination the other way around. Although the XRTT was a useful machine, its success was in the most part due to its most famous rider, Cal Rayborn, who often beat machines of greater power on this XRTT.

SPECIFICATIONS
1972 XRTT

- **ENGINE** Overhead-valve, V-twin
- **CAPACITY** 45cu. in. (750cc)
- **POWER OUTPUT** 90bhp @ 8,000rpm
- **TRANSMISSION** Four-speed, chain drive
- **FRAME** Tubular cradle
- **SUSPENSION** Telescopic front forks, swingarm rear
- **WEIGHT** 324lb (147kg)
- **TOP SPEED** 130mph (209km/h) (est.)

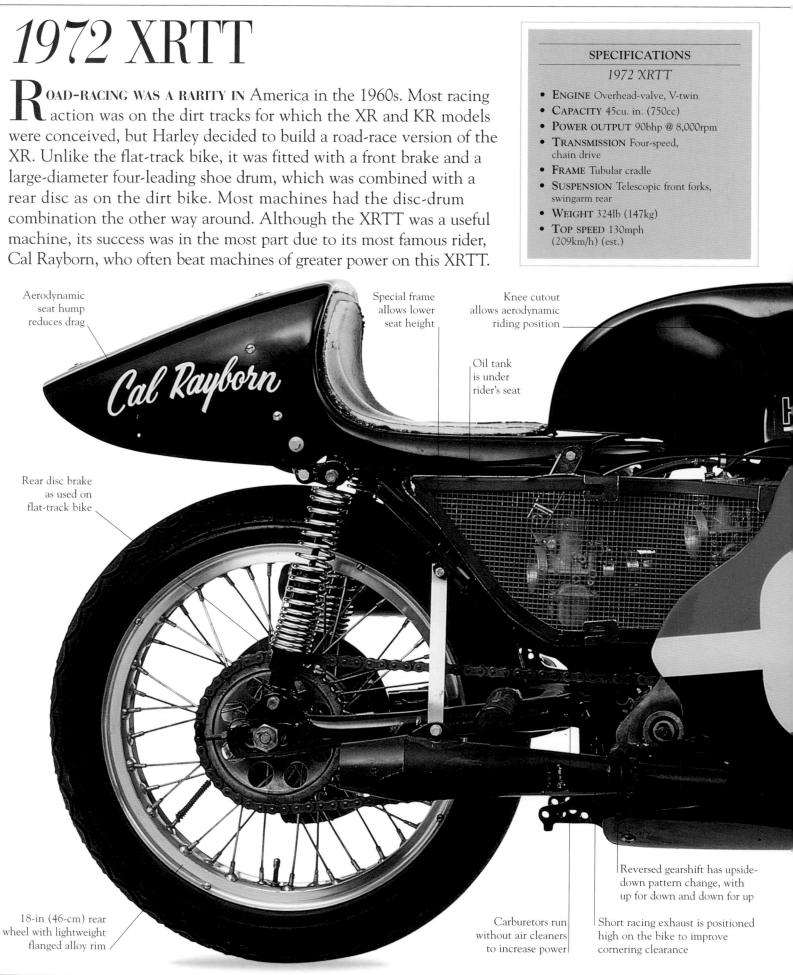

Aerodynamic seat hump reduces drag

Special frame allows lower seat height

Knee cutout allows aerodynamic riding position

Oil tank is under rider's seat

Rear disc brake as used on flat-track bike

Cal Rayborn

18-in (46-cm) rear wheel with lightweight flanged alloy rim

Carburetors run without air cleaners to increase power

Reversed gearshift has upside-down pattern change, with up for down and down for up

Short racing exhaust is positioned high on the bike to improve cornering clearance

1972 XRTT

Differences on the TT models from the dirt-track bikes included a modified frame, a larger capacity fuel tank, and an aerodynamic seat and fairing. Despite Cal Rayborn's notable efforts, these road-race versions of the XR did not have as much success as the dirt-track models.

Classic chassis
The XRTT chassis was derived from that used earlier on the K-series road racers (see pp.122–23). These frames continued to give service until the early 1980s, when they were used on Harley's "Battle of the Twins" racer.

Rider's chin-rest pad

Centrally mounted rev-counter is the bike's only instrument

Bikini fairing

Polished alloy fork yoke

Aerodynamic front fairing

Exhaust from rear cylinder

Underslung rear brake caliper

Front cylinder exhaust pipe

Brake cable

Ribbed pattern racing tire

Clip-on handlebars clamp directly onto the fork leg

Fairing painted in Harley-Davidson racing colors

Cerani four leading-shoe drum brake with huge ventilation slots to aid cooling

Tire is screwed to the rim for added security

1978 XLCR

I**N 1977 HARLEY INTRODUCED** a new variation of the Sportster. The XLCR—it can't have been an accident that it sounded like "excelsior" —was designed by Willie G. Davidson, and the CR stood for café racer. Harley's model was a blend of 1960s café racer—a name given to stripped, road-going hot rods used for blasting from bar to bar—with some of the styling cues of the XR flat-track racers. The frame and the exhaust pipes were new and would be used on the rest of the Sportster range the following year, but the engine was a stock XL1000. The really important stuff was the bodywork and the black finish. The XLCR looked great, but never sold in the numbers that were hoped for and was dropped after only two years.

SPECIFICATIONS
1978 XLCR
- **ENGINE** Overhead-valve, V-twin
- **CAPACITY** 61cu. in. (1000cc)
- **POWER OUTPUT** 55bhp
- **TRANSMISSION** Four-speed, chain drive
- **FRAME** Tubular cradle
- **SUSPENSION** Telescopic front forks, box-section swingarm rear
- **WEIGHT** 470lb (213kg)
- **TOP SPEED** 105mph (169km/h)

Fiberglass tailpiece/mudguard

Triangulated frame section derived from the XR750

Solo seat; a dual seat was also available in 1978

Battery housing cover

4-gallon (15-liter) fuel tank

Rear drive sprocket

Rear shock absorber mounting placed close to axle

Rear-set footrest and brake pedal linkage

Redesigned rear frame allows the oil tank (on other side of the battery) to be tucked in

Brake pedal

1978 XLCR

One of the reasons behind the bike's failure was that there was not enough stock to meet demand that had been generated by the prerelease publicity. Now, however, the small amount of numbers produced—about 3,200—means that this is something of a collector's item.

"Bar and shield" logo first used in 1910

Tinted windshield

A funked-up XL

Differences from the stock XL on which it was based included positioning the seat further backward and lowering the height of the handlebars. The rear section of the frame was derived from that used on the XR750 dirt-track bikes (see pp.128–29).

Drag-style straight handlebars

Headlight contained in fairing

Bikini fairing

Sculpted fuel tank

Rear shock absorber

Fold-up footrest

18-in (46-cm) rear wheel

Chromed suspension shroud

Shortened racing mudguard

Oil cooler

Wrinkle finish black engine paint

Speedometer cable

"Siamese" twin exhaust system

Exhaust retaining clamp

Morris 19-in (48-cm) seven-spoke alloy wheel

Kelsey-Hayes twin front disc brakes

1980 XR750

THE HARLEY-DAVIDSON **XR750** is the most successful competition bike ever produced, though for many it's more famous as the bike that jump hero Evel Knievel used for his stunts. The early bikes were introduced in 1970 with iron barrels and heads that failed miserably, so a revised alloy engined version of the bike was introduced two years later. For the first time on a production Harley V-twin, the rear cylinder had a forward-facing exhaust and rear-facing inlet port. The bike won the AMA Grand National Championship in its first year and, upgraded and improved over time, it is still winning races more than a quarter of a century later.

SPECIFICATIONS
1980 XR750

- **ENGINE** Overhead-valve, V-twin
- **CAPACITY** 45cu. in (748cc)
- **POWER OUTPUT** 90bhp @ 8,000rpm
- **TRANSMISSION** Four-speed, chain drive
- **FRAME** Tubular cradle
- **SUSPENSION** Telescopic front forks, swingarm rear
- **WEIGHT** 295lb (134kg)
- **TOP SPEED** 115mph (185km/h) (est.)

Lightweight fiberglass mudguard and seat base

Rear shock-absorber

Alloy oil tank is retained with springs

2½-gallon (9.5-liter) tank holds just enough fuel to last a race

Large air filter prevents dust from entering the engine

Dirt-track racing tire

Alloy drive sprocket can be easily changed to alter the gearing

Gear-change lever is seldom used once races are underway

Wide handlebars
provide extra
leverage

Fuel
filler-cap

Spot the difference
*The first XR750 was a comparative failure.
This bike is one of the iron-engined machines
produced from 1970–71 that used the same
Harley layout of carbs and
exhaust on the same side.*

Large racing seat

High ground clearance
is essential for all dirt-
track bikes

Steering
lock stop

*"The XR750 is a Harley-Davidson
racing classic, made famous by the
exploits of Evel Knievel and acknowledged
as one of the best dirt-track race
bikes ever produced."*

Standard Harley
frame was replaced by
most tuners seeking a
performance edge

Lightweight
telescopic
fork

Alloy
wheel rim

Alloy
cylinder
and head

Fins on the
cylinder barrel
help to disperse heat

Lightweight
front hub
with no brake

1980 XR750

Compare the engine to that of a Sportster
(*see pp.126–27*) and the origins of the bottom half
are obvious. However, the all-alloy top end, with rear
racing carburetors, was unique. The crankshaft was
different too and the 750cc capacity was achieved with
a bore of 3.125 inches and a stroke of 2.98 inches.

1984 XR1000

IT SEEMED OBVIOUS. Harley's XR750 (*see pp.128–29*) was cleaning up in dirt-track racing, so why not offer a limited run of road-going examples? In 1983 Harley produced the XR1000. They put the alloy heads and twin Dell'Orto carburetors from the XR750 onto the bottom half of an XL1000 engine. The engine itself was put into a standard XLX chassis. The result was an engine that put out 10 percent more power than the stock Sportster in standard trim, though many owners tuned the bike even more. The result was the fastest production bike Harley ever made, but buyers were not impressed—it looked almost identical to the cheapest bike in the line while costing a great deal more. The XR1000 didn't sell, but the bike is now a collector's piece.

SPECIFICATIONS
1984 XR1000

- **ENGINE** Overhead-valve, V-twin
- **CAPACITY** 61cu. in. (998cc)
- **POWER OUTPUT** 70bhp @ 6,000rpm
- **TRANSMISSION** Four-speed, chain drive
- **FRAME** Tubular cradle
- **SUSPENSION** Telescopic front forks, swingarm rear
- **WEIGHT** 470lb (213kg) (est.)
- **TOP SPEED** 120mph (193km/h) (est.)

Rear light

Alloy mudguard support

Oil tank

Non-standard Corbin seat; the XR was supplied with a solo saddle only

Classic Sportster fuel tank holds 2¼ gallons (8.3 liters)

Corbin

Passenger footrest

Air filter

16-in (41-cm) nine-spoke alloy rear wheel

Box section steel swingarm

Rear brake master cylinder

Rearview mirror

Low handlebars

Front brake master cylinder

Speedometer and rev-counter

Headlight eyebrow

Race success
In 1984, Gene Church won the Battle of the Twins race at Daytona on a heavily tuned 112bhp XR1000, retaining his title for the next two years.

High-level exhaust pipes mounted on the left-hand side

❝Successes on the race track brought the XR's potential to the public's attention, to the point where it is now seen as a classic Harley.❞

Telescopic front fork

Though this model is painted slat-gray, Harley's orange and black racing livery was optional for 1984

Slim mudguard and other components taken from the base XLX model

19-in (48-cm) nine-spoke alloy front wheel

Cast-iron cylinders are topped by alloy heads

Safety reflector

Generator mounting

New brake caliper design

10-in (25-cm) front brake discs

1984 XR1000

There was massive initial interest in the XR1000, but speed-hungry buyers expecting a new breed of Harley were disappointed. Only the few who bought the $1,000 bhp-doubling tuning kit saw the bike's real potential. Some people maintain that the XR1000 is the best bike Harley ever built.

1987 XLH883

HARLEY NEEDED HELP in the mid-1980s. It wanted to lure new customers who would resist the suspect reliability and high maintenance of a traditional Harley, but would still want the image. In addition, the company was required to meet strict new emission and noise laws. The end result was the "Evo" Sportster. It looked and sounded (once you'd taken the baffles out of the mufflers) just like an old Sportster, the difference being that the all-new engine was cheap and reliable. Introduced for 1986, the XLH had an 883cc block with alloy barrels and heads. Larger 1100cc and 1200cc units followed soon after, as did belt final-drive and a five-speed box. A competitively low price tag on this base model helped attract a whole new group of buyers who wanted to sample the Harley legend.

SPECIFICATIONS
1987 XLH883

- **ENGINE** Overhead-valve, V-twin
- **CAPACITY** 54cu. in. (883cc)
- **POWER OUTPUT** 49bhp @ 7,000rpm
- **TRANSMISSION** Five-speed, chain drive
- **FRAME** Tubular cradle
- **SUSPENSION** Telescopic front forks, swingarm rear
- **WEIGHT** 470lb (213kg)
- **TOP SPEED** 105mph (169km/h)

Aftermarket seat

License plate mounted above the rear light

Oil tank for dry sump lubrication system

Alloy rocker cover

Rear light

16-in (41-cm) rear wheel

Dual exhaust system with twin mufflers

1987 XLH883
The XLH helped pull Harley-Davidson out of trouble at a time when it was losing vast amounts of money. Its combination of quiet Evolution engine, reliability, classic design, and cheap price made it an immediate success.

Rearview mirror

No fancy details on the instrument dial, just a speedometer

Wide-bladed control lever

Traditional Sportster handlebars

2½-gallon (10-liter) "peanut" fuel tank

The 883cc engine capacity was a return to the Sportster's original 1957 engine size

Clean lines

As well as the impressive new engine, the bike's simplicity was one of its main selling points. The single instrument dial, plain paint job, and small tank all made for a neat-looking machine that found favor with tens of thousands of new Harley buyers.

Braced steering head

Alloy fork slider

Throttle cable

Narrow build and low seat height made the Sportster appealing to novice riders

Safety reflector

Single 11½-in (29-cm) front disc brake

New federal noise regulations were satisfied by adding a balance tube between the twin exhaust pipes

The large "ham can" air filter was first introduced on the Sportster in 1966 and still gets in the way of the rider's right leg

This model has traditional wire wheels, but cast spoked wheels were offered as an optional extra

Chrome wheel rim

1999 XL1200S

Though the Sportster engine has been through various guises and capacities over the years, it has always remained true to its original layout. In a sense, the XL1200S harks back to the XLCH Sportster of the late 1950s and early 1960s, a bare-boned bike intended for fast fun. A 1200cc version of the alloy Evolution engine (*see pp.144–45*) was introduced in 1988, and 11 years later it has received only detail changes. The Sportster Sport has upgraded suspension and improved power output over the basic model to justify its S designation. In addition, the engine is equipped with twin-plug heads and has increased compression and revised camshafts in comparison with the base model. There's no doubt that the XL1200S is a true return to form.

1999 XL1200S

When the 1200 Sport was originally released in 1996, it was the first road-going Harley to feature adjustable sporting suspension. Other additions such as the 13-spoke wheels and twin-plug heads turned this into Harley-Davidson's most adventurous model. Three years on, the minimalist 1999 bike is arguably the most stylish and competent Sportster that Harley has ever made.

Gas reservoir rear shock absorber is adjustable for pre-load and damping

Low profile seat

3⅓-gallon (12.5-liter) fuel tank larger than on previous Sportsters

Oil tank filler-cap

Taillight

16-in (41-cm) spoked alloy rear wheel

Large-diameter sports muffler

Belt drive pulley

Belt drive-pulley; belt final drive has been used on Sportsters since 1991

Brake master cylinder

Dual exhaust system

Five-speed gearbox

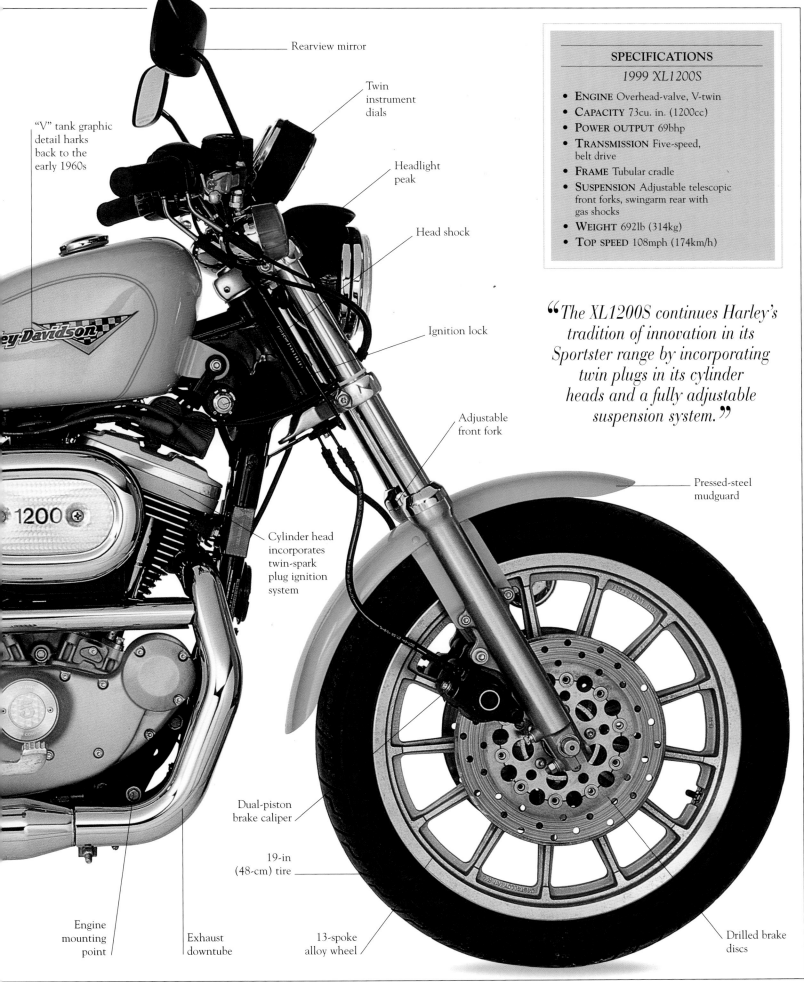

Rearview mirror

Twin instrument dials

"V" tank graphic detail harks back to the early 1960s

Headlight peak

Head shock

Ignition lock

Adjustable front fork

Cylinder head incorporates twin-spark plug ignition system

Pressed-steel mudguard

Dual-piston brake caliper

19-in (48-cm) tire

Engine mounting point

Exhaust downtube

13-spoke alloy wheel

Drilled brake discs

SPECIFICATIONS
1999 XL1200S

- **ENGINE** Overhead-valve, V-twin
- **CAPACITY** 73cu. in. (1200cc)
- **POWER OUTPUT** 69bhp
- **TRANSMISSION** Five-speed, belt drive
- **FRAME** Tubular cradle
- **SUSPENSION** Adjustable telescopic front forks, swingarm rear with gas shocks
- **WEIGHT** 692lb (314kg)
- **TOP SPEED** 108mph (174km/h)

"The XL1200S continues Harley's tradition of innovation in its Sportster range by incorporating twin plugs in its cylinder heads and a fully adjustable suspension system."

2003 XL883R

HARLEY'S LONG HISTORY OF flat-track racing success has never been a rich seam of inspiration for production models. The XLCR (*see pp.126–27*) and XR1000 (*see pp.130–31*) were both influenced by the all-conquering XR750 (*see pp.128–29*), but neither was a commercial success. In 2002 the factory tried again with the 883R. The new model was little more than a pared-down version of the base model Sportster fitted with black cowhorn handlebars and a two-into-one exhaust. Despite the limited modifications, it was an inspired addition to the range. Compared to many modern machines, the 883R's handling is unrefined, but it is a hugely involving bike to ride in town, or on country roads where its limited performance isn't an issue.

2003 XL883R

The six-model Sportster range is derived from one frame, two engines, a couple of different fuel tanks, and a variety of handlebar shapes and color schemes. Yet the different models can feel markedly different. The 883 is seen as an entry-level model, but the racetrack credibility of the orange and black livery—it is inspired by factory racers—helps make the R version attractive to flat-track fans.

3⅓-gallon (12.5-liter) fuel tank allows an uninterrupted ride of over 150 miles (240 km).

Single rear shock absorber is adjustable for preload only

Stepped dual seat provides a level of back support

Taillight

16-in (41-cm) rear wheel

Drive belt pulley

Large-diameter sports silencer

Oil filler-cap

Satin black oil tank

Curved two-into-one exhaust system

Fuel filler-cap

Short-stem rear-view mirror

Centrally mounted speedometer

"Bullet"-style indicator

29.6° steering head

Fork shroud

Safety reflector

Four-piston front brake caliper

Air filter

Chromed exhaust downtube

19-in (48-cm) front wheel

Twin drilled 11½-in (29-cm) brake discs

SPECIFICATIONS
2003 XL883R

- **ENGINE** Overhead-valve, V-twin
- **CAPACITY** 54cu. in. (883cc)
- **POWER OUTPUT** 45bhp @ 6,000rpm
- **TRANSMISSION** Five-speed, belt drive
- **FRAME** Tubular cradle
- **SUSPENSION** Telescopic front forks, swingarm rear
- **WEIGHT** 540lb (245kg) (dry, claimed)
- **TOP SPEED** 103mph (166km/h)

"In a world bored by retro-kitsch, this is the real deal—not some new-VW Beetle pastiche of an old design. This is a new design."

2010 XR1200X

HARLEY'S **XR750** is an iconic race bike that has dominated American dirt track racing since 1972. However, Harley were never able to effectively translate the image of the XR to a road bike. They attempted this in 2008 when the XR1200, a more aggressive version of the Sportster, was introduced. The new bike was originally intended for European markets, though it was subsequently sold in America too. Based on the 1200 Sportster, the XR1200 featured a tuned engine, uprated suspension, improved brakes, and sports tires.

SPECIFICATIONS
2010 XR1200X

- **ENGINE** Overhead-valve, V-twin
- **CAPACITY** 1250cc
- **POWER OUTPUT** 90bhp @ 7,000rpm
- **TRANSMISSION** Five-speed, belt drive
- **FRAME** Tubular cradle
- **SUSPENSION** Telescopic fork front, swingarm rear
- **WEIGHT** 412lb (187kg)
- **TOP SPEED** 125mph (201km/h)

Rack and backrest are factory options

Seat echos styling of XR750

Oil-cooled cylinder heads

Oil tank

Rear suspension units have adjustable damping

Twin exhausts echo styling of XR750 racers

17-inch (43-cm) rear wheel

Alloy swingarms

Raised footrests give improved cornering clearance

Mirrors have unobtrusive black finish

3½-gallon (13.25-liter) fuel tank

XR1200X

For 2010 an X was added to the model number of the XR1200. The bike was fitted with better quality suspension and offered with a black paint finish. Showa multi-adjustable suspension featured remote reservoir "piggy back" shock absorbers and "upside down" forks for greater chassis rigidity.

Analog revcounter is centrally mounted, with smaller digital speedo to one side

"In the Sportster's half century of history, the fastest and best-handling version of the model that Harley ever built was the XR1200."

Showa forks

Twin disc brakes are fitted

Vibration-isolating front engine mounting

Two-into-one-into-two exhaust system

Four-piston front brake calliper

18-inch (46-cm) front wheel

─ *CHAPTER SEVEN* ─

RECENT BIG-TWINS

1984–2010

1999 FAT BOY

IN THE LAST QUARTER OF THE 20TH CENTURY,
motorcycle designers were faced with a problem.
How do you meet strict new noise and emissions
regulations without hindering performance?
In addition, Harley-Davidson needed to appeal to
buyers who liked the idea of a bike that looked as
though it was designed three decades earlier, but
performed like a modern machine. The solution
was the 1984 Evolution and the 1998 Twin Cam.

NEW LEASE OF LIFE
*Harley's new improved engines may have looked like the old
ones, but they came with increased power and, more
importantly to enthusiasts, improved reliability.*

1988 FLHS Electra Glide

THE FLHS WAS A NEW DERIVATIVE of the classic Electra Glide introduced in 1987. At the heart of the bike was the 80cu. in. Evolution engine and five-speed transmission that was first released in 1984. Also present on the bike was belt final-drive, which became common on Harleys from 1985. The FLHS was a return to the traditions of the earliest Electra Glides and was sold with no top box and a removable windshield. Though the "Sport" tag might have been stretching a point, the trimmed-down model certainly felt more agile than the fully encumbered standard version. It also had the advantage of being considerably cheaper.

1988 FLHS ELECTRA GLIDE
The FLHS designation had been around for a few years, with the first example introduced for 1977 as a one-year-only limited edition model. Harley then decided to withdraw a sport edition of the Electra Glide for three years before returning as the 1980 FLHS. This was carried through until the end of the Shovelhead engine in 1984 and a break of a couple of years ensued before the FLHS returned as a sport version of the new Evolution Electra Glide in 1987.

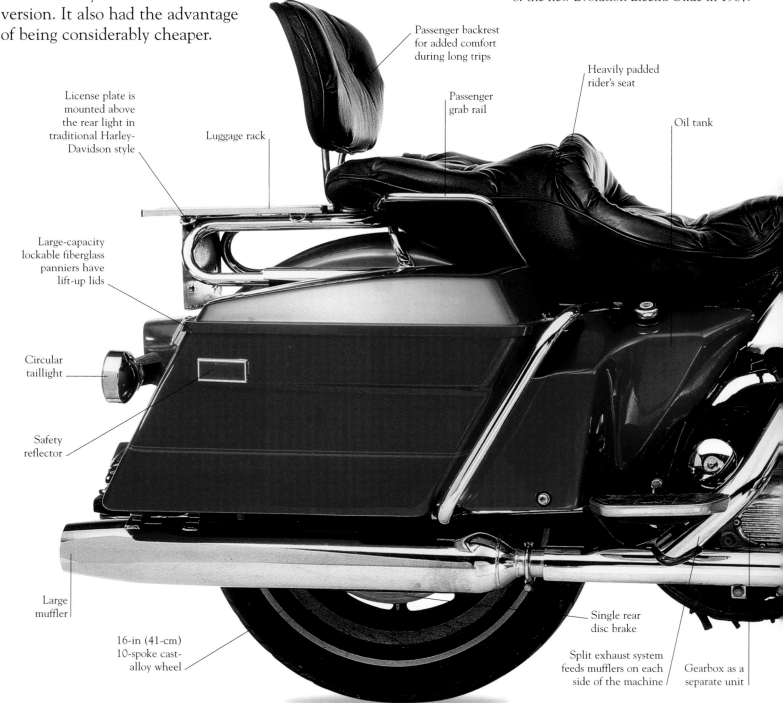

Passenger backrest for added comfort during long trips

Heavily padded rider's seat

Passenger grab rail

Oil tank

License plate is mounted above the rear light in traditional Harley-Davidson style

Luggage rack

Large-capacity lockable fiberglass panniers have lift-up lids

Circular taillight

Safety reflector

Large muffler

16-in (41-cm) 10-spoke cast-alloy wheel

Single rear disc brake

Split exhaust system feeds mufflers on each side of the machine

Gearbox as a separate unit

Rearview mirror

Brake fluid reservoir

Rubber handlebar grip

5-gallon (19-liter) fuel tank

Bar and shield tank logo

Detachable windshield is one of the elements that makes this a Sport version

Chrome headlight nacelle

Fog lamp

Self-canceling turn signals were new for 1988

Pressed-steel front mudguard

Evolution cylinders

Air filter

Footboard

Crash bar

Dual front disc brakes

18-in (46-cm) tire

Alloy fork slider

SPECIFICATIONS
1988 FLHS Electra Glide
- **ENGINE** Overhead-valve, V-twin
- **CAPACITY** 80cu. in. (1312cc)
- **POWER OUTPUT** 55bhp @ 7,200rpm
- **TRANSMISSION** Five-speed, belt drive
- **FRAME** Tubular cradle
- **SUSPENSION** Telescopic front forks, twin rear shocks, swingarm rear
- **WEIGHT** 692lb (314kg)
- **TOP SPEED** 100mph (161km/h)

"The Electra Glide is the most enigmatic of all Harley-Davidsons, the king of touring bikes and the undisputed leader of the pack."

The Evolution

ONE MILLION CUSTOMERS can't be wrong. That is how many Evolution engines were built in its 15 years of production from 1984 to 1999. The engine followed the traditional Harley format of a 45° air-cooled V-twin with two overhead valves per cylinder, but it was more reliable and more efficient than its predecessor, the venerable Shovelhead. The Evolution gave Harley the chance to get on with developing new bikes and new markets without having to worry about the engine. While the standard engine produced a restrained 69bhp, many owners were happy to make the bike noisier and more powerful.

FLHS Electra Glide
The Evolution was used to power an increasingly broad range of machines that covered everything from this traditional Electra Glide to the FLSTF Fat Boy.

Three-piece alloy rocker cover conceals hydraulic tappets

Alloy cylinders ran more coolly than the Shovelhead's iron ones and so were more efficient

Air filter mounting bolt

Carburetor

Circular air filter

Cylinders have a 3.5 x 4.25in bore and stroke

Exhaust port with studs for exhaust pipe fitting

Polished pushrod tubes are used to return oil from the head to the crankcase; oil is forced up to the heads through the pushrods themselves

Hydraulic tappets are situated in the blocks at the base of the pushrod tubes

INSIDE THE EVOLUTION

The Evolution part of the engine was really in the new cylinders, heads, ignition, and carburation systems, which were attached to a lower end based on the last of the Shovelheads. Alloy cylinders, improved combustion-chamber shape, and flat-topped pistons made the Evo run more coolly and more efficiently than the Shovel. An improved carburetor and a new "V-Fire III" electronic ignition system also helped.

A high-quality touring bike

The Evolution arrived on the scene at a time when touring was a well-established part of motorcycle culture. Increased reliability, more miles to the gallon, and reduced weight over the Shovelhead meant the Evo was an ideal power plant for all types of touring.

Crankshaft and bottom end are the same as on the Shovelhead

"Overnight the Evolution motor transformed Harley's fortunes. It came at just the right time, worked faultlessly, and really saved the Hog's bacon."

JOHN WARR
(HARLEY DEALER)

Oil pump is in the traditional position at the rear of the timing case

Camshaft is mounted beneath the center of the "V"

The alternator cover also hides the trigger for the electronic ignition

"This was the engine that secured Harley's future. It killed the poor reputation that had developed during the difficult AMF years."

JOHN WARR
(HARLEY DEALER)

THE COMPETITION

• 1987 HONDA GL1500 GOLDWING •
In the six-cylinder GoldWing, Honda had a fierce competitor to Harley's Evo bikes. More so as the 'Wing was made in Ohio and wore its "Made in America" badge almost as proudly as the Milwaukee machines.

2000s | 1990s | 1980s | 1970s | 1960s | 1950s | 1940s | 1930s | 1920s | 1910s | 1900s

1989 FLTC Tour Glide

BY CONTRAST WITH THE trimmed-down FLHS (*see pp.142–43*), the FLTC represented the "fully loaded" end of the Electra Glide line. Harley understood that it could offer a variety of models based on similar ingredients to suit the needs of all its customers, and in the case of the FLTC—first introduced for the 1984 model year—the need was for complete luxury. The fairing has twin headlamps and is frame-mounted to improve the bike's handling. The special two-tone paint finish was used on the "Classic" (FLTC) model, but there was also a slightly cheaper FLT version with a single-color paint finish.

SPECIFICATIONS

1989 FLTC Tour Glide

- **ENGINE** Overhead-valve, V-twin
- **CAPACITY** 80cu. in. (1312cc)
- **POWER OUTPUT** 58bhp @ 7,200 rpm
- **TRANSMISSION** Five-speed, belt drive
- **FRAME** Tubular cradle
- **SUSPENSION** Telescopic front forks, swingarm rear
- **WEIGHT** 732lb (332kg)
- **TOP SPEED** 110mph (177km/h) (est.)

Passenger armrest

Top box adds to the already expansive luggage capacity

Massive twin headlights allow safe night riding

Additional side reflectors

Footboard

Self-canceling indicator

Muffler

Rear disc brake

Oil tank

White striping on 18-in (46-cm) tire

Sidecar companion
The FLTC was an ideal bike for a sidecar and buyers could order one from Harley. Numbers were never high, however, with 15 produced in 1989 out of a total FLTC production figure of 603.

Rocker cover badge
The sentiment expressed on this replacement for the standard air filter—and also inscribed on the passenger footboard—is typical of the die-hard Harley-Davidson fanatic.

Smoked windshield creates a calm pocket of air for a relaxed ride

40-watt stereo system; a CB radio came with the "Ultra" package

Handlebar-mounted controls can adjust the stereo volume

Two-tone paint finish

Deep-cushioned, solidly mounted seat

"The FLTC was the top-of-the-line Electra Glide, and included features such as a 40-watt music system, twin headlights, and all the luggage space you could want."

1989 FLTC TOUR GLIDE CLASSIC
The Tour Glide Classic was based on the FLT, which first appeared in 1980. This significant new model had a revised frame which incorporated a vibration-isolated engine and a box-section spine, these features being adopted by all Electra Glides from the mid-1980s. The FLTC weighed over half a ton when carrying two riders and a full load.

Fiberglass fairing mounted to the frame rather than the forks to improve stability

Highway peg mounted on safety bar where rider can rest feet

Safety reflector

Folded-up passenger footboard

Two-into-one exhaust system

Five-speed gearbox

Old-style footboards add to rider comfort

Chrome mudguard trim

Ten-spoke alloy wheel

1989 FXR Super Glide

HARLEY'S SUPER GLIDE WAS revised in 1982 with the introduction of the FXR, the base model in an expanded line of Super Glide-derived bikes. The FXR got an 80cu. in. engine, a new frame, and many other changes, but it vanished in 1984 when the new Evolution engine appeared. It re-emerged in 1986 in a similar guise and continued until 1994, when the Dyna-framed Super Glide (*see pp.154–55*) replaced it. Though the bike has been modified over the years, the original Super Glide concept of a big-twin engine in a cruiser/custom chassis has remained constant. A good idea is timeless.

1989 FXR SUPER GLIDE
Harley offered a range of factory-equipped optional extras for the FXR that included twin front disc-brakes, a solo seat, Sportster fuel tanks, and spoked wheels. A tuning kit was also available that boosted power output to over 80bhp. However, this example has been given a number of non-factory extras by its owner; note the handlebar tassles, the "slash-cut" mufflers, the windshield, and the panniers.

> *"The FXR was a continuation of Harley's successful Super Glide theme of placing the biggest V-twin engine of the line in a cruiser chassis."*

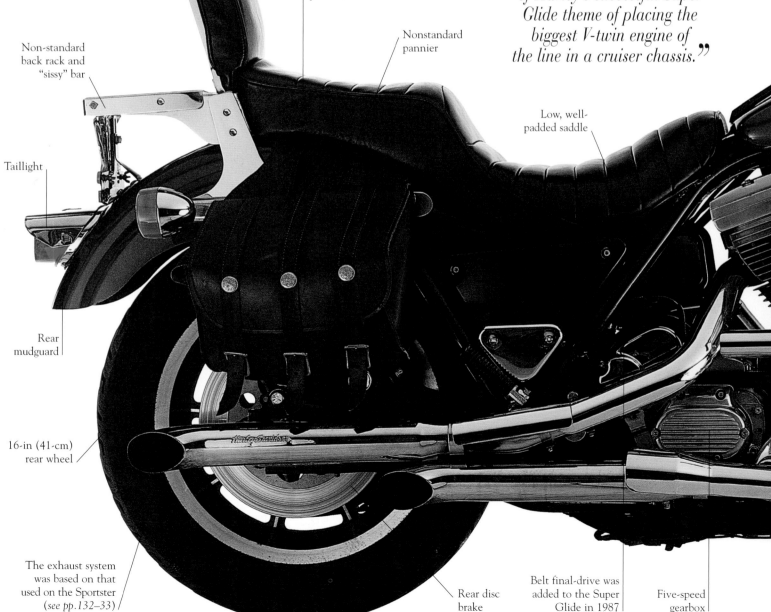

Passenger backrest

"King and queen" seat

Nonstandard pannier

Non-standard back rack and "sissy" bar

Low, well-padded saddle

Taillight

Rear mudguard

16-in (41-cm) rear wheel

The exhaust system was based on that used on the Sportster (*see pp.132–33*)

Rear disc brake

Belt final-drive was added to the Super Glide in 1987

Five-speed gearbox

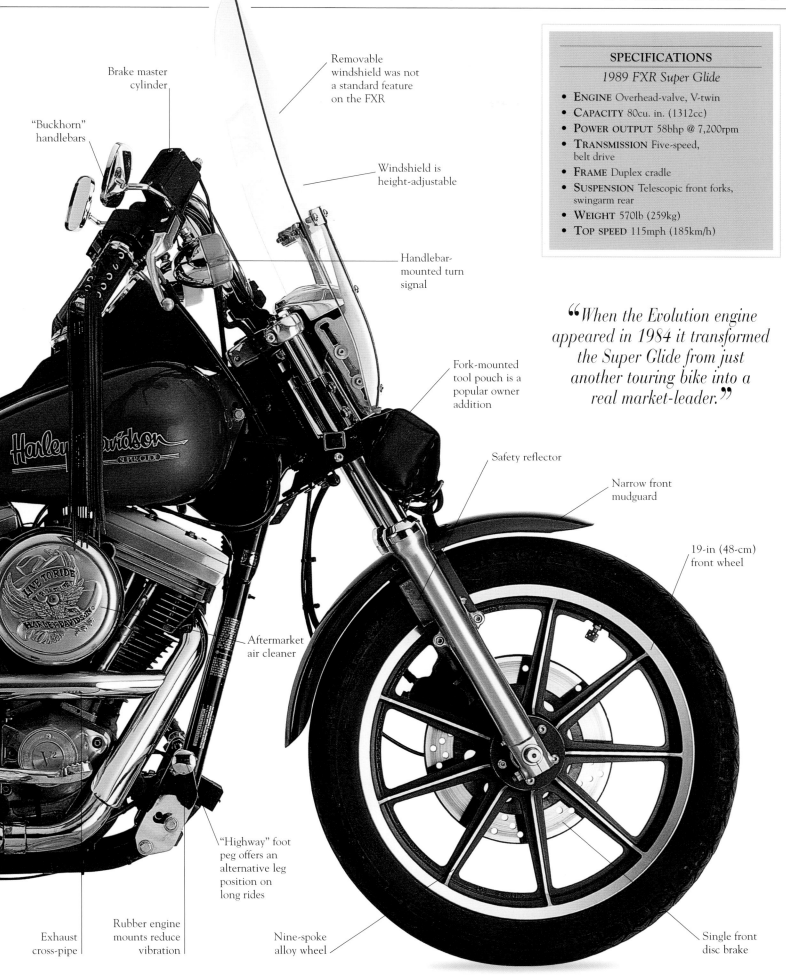

Brake master cylinder

"Buckhorn" handlebars

Removable windshield was not a standard feature on the FXR

Windshield is height-adjustable

Handlebar-mounted turn signal

Fork-mounted tool pouch is a popular owner addition

Safety reflector

Narrow front mudguard

Aftermarket air cleaner

"Highway" foot peg offers an alternative leg position on long rides

Exhaust cross-pipe

Rubber engine mounts reduce vibration

Nine-spoke alloy wheel

19-in (48-cm) front wheel

Single front disc brake

SPECIFICATIONS
1989 FXR Super Glide

- **ENGINE** Overhead-valve, V-twin
- **CAPACITY** 80cu. in. (1312cc)
- **POWER OUTPUT** 58bhp @ 7,200rpm
- **TRANSMISSION** Five-speed, belt drive
- **FRAME** Duplex cradle
- **SUSPENSION** Telescopic front forks, swingarm rear
- **WEIGHT** 570lb (259kg)
- **TOP SPEED** 115mph (185km/h)

"When the Evolution engine appeared in 1984 it transformed the Super Glide from just another touring bike into a real market-leader."

1997 FLHRI Road King

BY THE MID-1990S IT WAS QUITE obvious that there wasn't going to be any significant "new idea" which had a Harley-Davidson badge on the tank. Harley knew what its customers wanted, and it knew what it was good at building. Hence the regular reappearance of old ideas such as the Road King, a middleweight tourer first introduced for the 1995 model year. This was a return to traditional values of the Electra Glide in much the same way that the FLHS (*see pp.142–43*) had been a decade earlier. The Road King combined improvements to the Harley package—like electronic sequential port fuel injection on the FLHRI—with the looks of the traditional Electra Glide such as whitewall tires, spoked wheels, leather saddlebags, and plenty of chrome. And with a price tag of about $15,000, Harley-Davidson now had a bike to give the big Japanese tourers a run for their money.

SPECIFICATIONS
1997 FLHRI Road King

- **ENGINE** Overhead-valve, fuel-injected V-twin
- **CAPACITY** 80cu. in. (1312cc)
- **POWER OUTPUT** 69bhp
- **TRANSMISSION** Five-speed, belt drive
- **FRAME** Tubular cradle
- **SUSPENSION** Telescopic front forks, swingarm rear
- **WEIGHT** 692lb (314kg)
- **TOP SPEED** 96mph (155km/h)

1997 FLHRI ROAD KING
The Evolution engine had been a success since its introduction in 1984, but the Weber fuel-injection option available on the FLHRI transformed it into a different beast again. Economic on fuel, easy to start, and lower exhaust emissions all added to a new and improved riding experience.

Removable pillion seat

Removable saddlebag is made from leather covering a hard shell

Rear turn signal

Oil tank has 1-gallon (3.8-liter) capacity

Large muffler

Whitewall tire

Wire-spoked wheel

Single rear disc brake

Swingarm pivot

Passenger footboard (folded up)

Quick-detach
windshield

Electronic
starter button

Wide-bladed
handlebar lever

5-gallon
(19-liter)
fuel tank

Riding stance
*The handlebars were positioned high and
wide, which was good for short- to middle-
distance touring but not so practical for
longer trips. The bike's narrow profile
and low center of
gravity did, however,
provide enjoyable riding
on twisting roads.*

Passing lights are
now a traditional
feature of the
touring Harley

Footboard

Traditional
valanced
mudguard

Road King
mudguard
script

Air filter cover
boasts that the
bike is fitted with
fuel injection

Footboard

Chrome trim

80cu. in. Evolution engine is
positioned in rubber-mounting
system to reduce vibration

Dual front
disc brakes

1999 FLSTF Fat Boy

HARDTAILS WERE MOTORCYCLES without rear suspension, a customizing trend intended to give the rear of a machine a cleaner look. In another example of Harley-Davidson being influenced by the way its bikes were customized, the company introduced the "Softail" in 1984. Harley wanted the clean look but didn't want to inflict the discomfort of a solid chassis onto its riders, so the bike had the look of a hardtail but with rear suspension units hidden under the engine. The FLSTF Fat Boy, launched in 1990, was a further variation on this idea, with solid disc wheels and a unique exhaust system contributing to the Fat Boy look. The FLSTS Heritage Springer went even further by using the Springer forks that Harley had discontinued in 1949 when it introduced the Hydra-Glide (*see pp.84–85*).

SPECIFICATIONS

1999 FLSTF Fat Boy

- **ENGINE** Overhead-valve, V-twin
- **CAPACITY** 80cu. in. (1312cc)
- **POWER OUTPUT** 83bhp
- **TRANSMISSION** Five-speed, belt drive
- **FRAME** Tubular cradle
- **SUSPENSION** Telescopic front forks, swingarm rear
- **WEIGHT** 598lb (271.5kg)
- **TOP SPEED** 120mph (193km/h)

1999 FLSTF FAT BOY
The solid disc wheels are the Fat Boy's most unusual feature, but solid covers for spoked wheels have been seen on Harleys at various times in the past, normally as an aftermarket accessory. The distinctive headlamp and fork panel was originally on the Hydra-Glide (*see pp.84–85*).

Rear mudguard is almost identical in style to that on the Hydra-Glide

Taillight

Ultra-low contoured seat is just 26½in (67.3cm) above road level

Rear disc brake

Softail swingarm has concealed suspension units

Low muffler hides the horizontally mounted shock absorber

Passenger footrest

Chromed oil tank

Wing mirror

Handlebar-mounted turn signal

Front brake lever

Wide-spaced handlebars

Electric starter

Turn signal

4⅕-gallon (15.9-liter) fuel tank

" *The Fat Boy's unique styling has made it the best-selling Harley of the 1990s and the perfect partner for Arnold Schwarzenegger in the 1991 film Terminator II.* "

Footboard

Instrument console

Chrome headlight

Clean front
Placing the speedometer and warning lights in the dash on the fuel tank gives the Fat Boy a very uncluttered front, which is part of its appeal.

Headlight is mounted to a metal panel on the forks as on the 1948 FL

Two-tone Lazer Red and black paint finish

Metal shrouds conceal the fork legs on the FL-style front end

Safety reflector

Footboard

Dual exhaust system

Evolution engine was replaced by the Twin Cam (*see pp.156–57*) on year 2000 Softail models

16-in (41-cm) alloy disc wheel

Dunlop Elite "fat" tire

1999 FXDX Super Glide

FOR THE NEW MILLENNIUM, Harley figured that it needed a new, more powerful, and more purposeful engine to replace the Evolution power unit which had provided such good service since 1984. This being Harley, it didn't want too much change all at once, so the new Twin Cam engine (*see pp.156–57*) could be, and was, slotted straight into the existing big-twin chassis. The new unit made its debut in 1998 on selected 1999 models, including the Super Glide Sport. Harley claimed a hefty 24-percent power increase—to a still unspectacular 68bhp—for the new engine, largely as a result of an increase in revs and an increase in capacity to 88cu. in. However, the new power enabled the Super Glide Sport to hit solid three-figure top speeds, and sustain them—an ability uncharacteristic for a Harley.

SPECIFICATIONS
1999 FXDX Super Glide

- **ENGINE** Overhead-valve, V-twin
- **CAPACITY** 88cu. in. (1450cc)
- **POWER OUTPUT** 68bhp (est.)
- **TRANSMISSION** Five-speed, belt drive
- **FRAME** Tubular cradle
- **SUSPENSION** Telescopic front forks, swingarm rear
- **WEIGHT** 615lb (279kg)
- **TOP SPEED** 110mph (177km/h)

1999 FXDX SUPER GLIDE SPORT
Harley first used a black engine finish on the XLCR (*see pp.126–27*) back in 1977, and although a black finish on engines and exhausts helps to dissipate heat, its popularity is more for cosmetic reasons. The matt finish on the FXD gives the engine a mean and purposeful look.

Sport version has a slimmer seat than the base FXD model

Battery case

Taillight

16-in (41-cm) wire-spoked wheel with Dunlop Elite tire

Ignition and parking light switch

Single rear disc brake

Nonstandard exhaust pipes release more power and noise

Wing mirror

Lower, black
handlebars
fitted on this
Sport model

Electronic
speedometer and
rev-counter

Handlebar-
mounted turn
signal

Fuel filler-cap

Fuel gauge

Chrome trimmed
headlight

5-gallon
(18.57-liter)
fuel tank

28° steering head

Steering
lock

Safety
reflector

Front brake
caliper

Engine
mounting point

88cu. in. Twin Cam
engine with wrinkle
black paint finish

Chromed
exhaust
downtube

Chrome
wheel rim

Twin front
disc brakes

Front brake lever

Essential components
*The Super Glide is a no-
flab motorcycle unburdened
with unnecessary features.
The front aspect is limited
to the essentials of the bike
and no more.*

Small-
diameter
headlight

Fork yoke

Fork
shroud

Footrest

Buyers could choose from
Aztec Orange Pearl,
Diamond Ice Pearl, or
Vivid Black paint finishes

Lightweight
front mudguard

The Twin Cam

TRADITION DICTATED THAT Harley engineers would ignore overhead camshafts, multi-valve cylinder heads, and other innovations when designing a new engine for the next millennium. And so it turned out. Though sophisticated fuel-injection and ignition systems provide a nod to the future, the 1999 Twin Cam still relies on the trusty 45° overhead-valve layout. The Twin Cam was a cleaner, more integrated design than the Evolution engine that preceded it (*see pp 144–45*) and it provided a platform for future development; engine capacity grew from 88cu.in. to 96cu.in. and then 103cu.in.

Connector

Power cable

Chain-driven efficiency
Removing the cam cover allows a view of the chain-drive to the cams. Replacing the gear-drive used on earlier engines reduced noise and manufacturing costs.

The Twin Cam was put through over 2½ million miles (4 million test kilometers) before its release

Exhaust port with stud for exhaust pipe fitting

Oval air-filter cover conceals the fuel-injection system

TWIN CAM 88™

Two-piece alloy rocker cover

New aluminum cylinders have a shorter stroke and longer bore than the Evolution

Exhaust port

Polished pushrod tubes

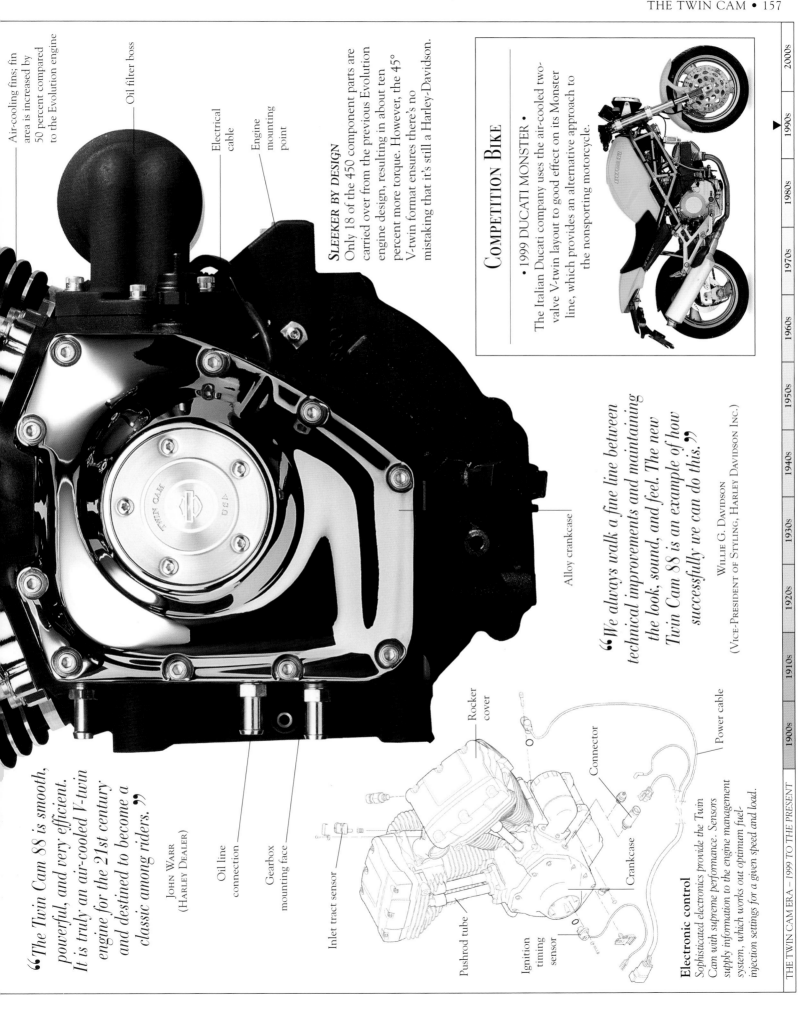

Air-cooling fins; fin area is increased by 50 percent compared to the Evolution engine

Oil filter boss

Electrical cable

Engine mounting point

Sleeker by Design
Only 18 of the 450 component parts are carried over from the previous Evolution engine design, resulting in about ten percent more torque. However, the 45° V-twin format ensures there's no mistaking that it's still a Harley-Davidson.

Competition Bike
• 1999 DUCATI MONSTER •
The Italian Ducati company uses the air-cooled two-valve V-twin layout to good effect on its Monster line, which provides an alternative approach to the nonsporting motorcycle.

"The Twin Cam 88 is smooth, powerful, and very efficient. It is truly an air-cooled V-twin engine for the 21st century and destined to become a classic among riders."
JOHN WARR
(HARLEY DEALER)

Oil line connection

Gearbox mounting face

Inlet tract sensor

Alloy crankcase

"We always walk a fine line between technical improvements and maintaining the look, sound, and feel. The new Twin Cam 88 is an example of how successfully we can do this."
WILLIE G. DAVIDSON
(VICE-PRESIDENT OF STYLING, HARLEY DAVIDSON INC.)

Rocker cover

Connector

Power cable

Crankcase

Pushrod tube

Ignition timing sensor

Electronic control
Sophisticated electronics provide the Twin Cam with supreme perfomance. Sensors supply information to the engine management system, which works out optimum fuel-injection settings for a given speed and load.

THE TWIN CAM ERA – 1999 TO THE PRESENT

2000s 1990s 1980s 1970s 1960s 1950s 1940s 1930s 1920s 1910s 1900s

1999 FXDWG Wide Glide

THE WIDE GLIDE IS A chopper-style machine inspired by the bikes featured in the film *Easy Rider*. With a huge customizing culture in the US, the Wide Glide was intended to give buyers the chopper look straight from the Harley-Davidson factory. The Wide Glide was an immediate hit when it first appeared in 1980—the widened fork yokes giving the bike its name—and soon became an established part of the Harley line. The Super Glide Dyna chassis is equipped with custom features that all come as standard, and for 1999 the bike incorporated the new 88cu. in. Twin Cam engine (*see pp.156–57*).

SPECIFICATIONS
1999 FXDWG Wide Glide

- **ENGINE** Overhead-valve, V-twin
- **CAPACITY** 88cu. in. (1450cc)
- **POWER OUTPUT** 79bhp
- **TRANSMISSION** Five-speed, belt drive
- **FRAME** Tubular cradle
- **SUSPENSION** Telescopic front forks, swingarm rear
- **WEIGHT** 598lb (271.5kg)
- **TOP SPEED** 120mph (193km/h)

1999 FXDWG WIDE GLIDE
The Dyna frame has a unique, computer-designed engine mounting system that uses flexible mountings and clever engineering to further reduce the effects of vibration on the motorcycle. The backbone of the frame is a rectangular section which is welded to a cast steering head.

Padded passenger backrest

"King and queen" seat with the pillion higher than the rider was another custom feature adopted by the factory

"Bobbed" rear mudguard

Rear turn signal

Seat height is a modest 26¼in (67.95cm)

16-in (41-cm) wire-spoked rear wheel with chrome rim

"Staggered shorty duals" exhaust system

Chrome battery box

Five-speed gearbox

"The Wide Glide has been updated ever since its introduction in 1980 and is now a true mix of classic custom styling with the most up-to-date Harley technology."

High-rise "Ape-hanger" handlebars

Handlebar controls feature wide-bladed levers and Harley's unique self-canceling turn signals

The "sissy bar," or passenger backrest, was intended to reassure nervous passengers

Chromed speedometer binnacle

Custom-style chrome headlight with single mounting point

Made in the USA badging
Harley customers around the world were getting a real piece of Americana when they bought their motorcycle. The company were happy to reassure them that they had made the right decision.

Widened fork yokes give the bike its name

Optional two-tone color scheme

21-in (53-cm) front wheel with skinny tire

Safety reflector

Forward-mounted foot controls

"Dyna" frame with vibration-isolating engine mountings

Chrome air filter displays the bike's 88cu. in. capacity

Single 11½-in (29-cm) front disc brake

Dunlop Elite tire

2003 FXSTB Night Train

INTRODUCED IN 1997, this variant of the Softail frame featured a pared-down custom look, a low profile seat, and flat, drag-racing handlebars. Originally fitted with the 1340cc Evo engine, the Night Train was given the 1450cc Twin Cam engine for the 2000 model year. Harley-Davidson was quick to notice new trends among motorcycle customizers; the model reflected a move away from candy paint finishes and exuberant use of chrome to a simpler appearance and predominantly black finish. The Night Train represents typical Harley policy of repackaging a variety of components in different ways to create a large number of models around one engine, three types of frames, and a raft of wheels, fuel tanks, handlebars, and seats.

SPECIFICATIONS
2003 FXSTB Night Train

- **ENGINE** Overhead-valve, V-twin
- **CAPACITY** 88 cu. in. (1450cc)
- **POWER OUTPUT** 69bhp (est.)
- **TRANSMISSION** Five-speed, belt drive
- **FRAME** Tubular cradle
- **SUSPENSION** Telescopic front forks, swingarm rear
- **WEIGHT** 673lb (305kg)
- **TOP SPEED** 120mph (193km/h)

2003 FXSTB NIGHT TRAIN
Riding a Harley is an experience. "You roll on the throttle and hear the engine roar," claims the marketing material for the Night Train. "Before long, the same primal urge that made cavemen go out and hunt woolly mammoths overtakes you." The reality may be less intriguing, but it is certainly an enjoyable ride.

Taillight

Low-profile seat

Rear disc brake

16-in (41-cm) solid rear wheel

"Staggered shorty duals" exhaust system

Softail swingarm unit

Passenger footrest

Twin Cam engine

Classic styling
The slim rear view and considerable length of the Night Train are classic Harley-Davidson styling features. By mounting the instruments in a tank-top console, the handlebars are kept clear to give a better view of the small, custom-style chrome headlight.

Wing mirror

5¼-gallon (19.7-liter) fuel tank

"Riser" handlebar mountings

Central instrument console incorporates speedometer, warning lights, and ignition switch

Flat, drag-racing handlebars

Classic "bobber"-style rear mudguard

Small, custom-style headlight

Turn signal

Chrome silencer

Front brake caliper

Lightweight front mudguard

Hydraulic brake hose

Forward control rear brake pedal

The engine's crinkle finish contrasts with the deep gloss black on the cycle parts

21-in (53-cm) spoked front wheel

2010 FLHR Road King

THE FL RANGE OF MACHINES can be traced all the way back to the original FL of 1941, and the current Road King plays on that heritage while also being a capable touring motorcycle for the 21st century. A ride-by-wire throttle, combined with digital fuel injection, constantly balances performance, economy, and emissions. At the same time, chassis improvements—a stiffer chassis with better suspension quality and improved geometry—means that the bike is easier to maneuver, and with a bike as big and heavy as this one, that is a really important consideration.

FLHR ROAD KING

In 2009 the Road King and its more luxurious stablemate, the Electra Glide, were given all-new chassis, resulting in better handling and superior ride quality. Improved engine mountings minimized vibration without losing the looks and riding character of the previous models, while better brakes were a welcome addition, and an anti-lock braking system was also an option.

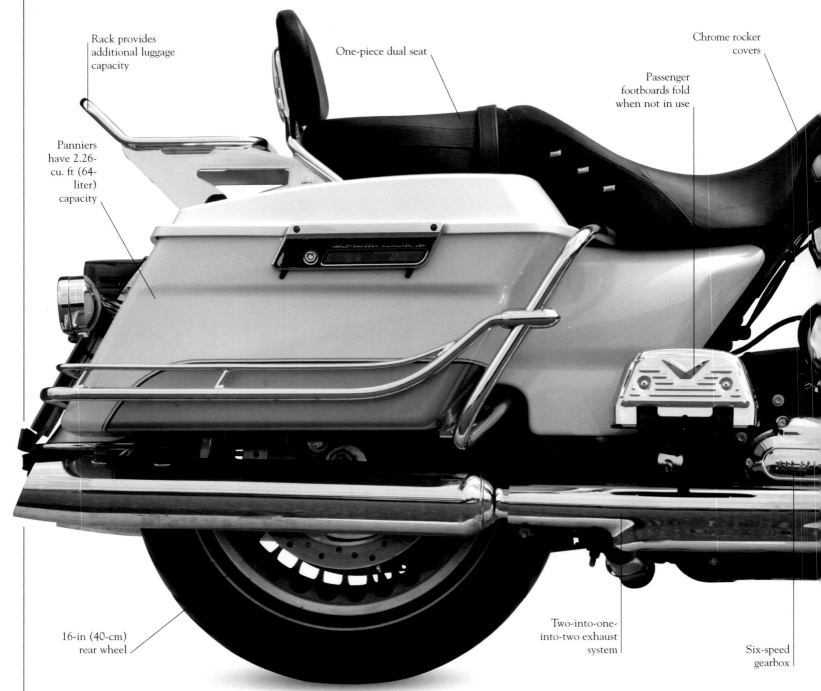

Rack provides additional luggage capacity

One-piece dual seat

Chrome rocker covers

Passenger footboards fold when not in use

Panniers have 2.26-cu. ft (64-liter) capacity

16-in (40-cm) rear wheel

Two-into-one-into-two exhaust system

Six-speed gearbox

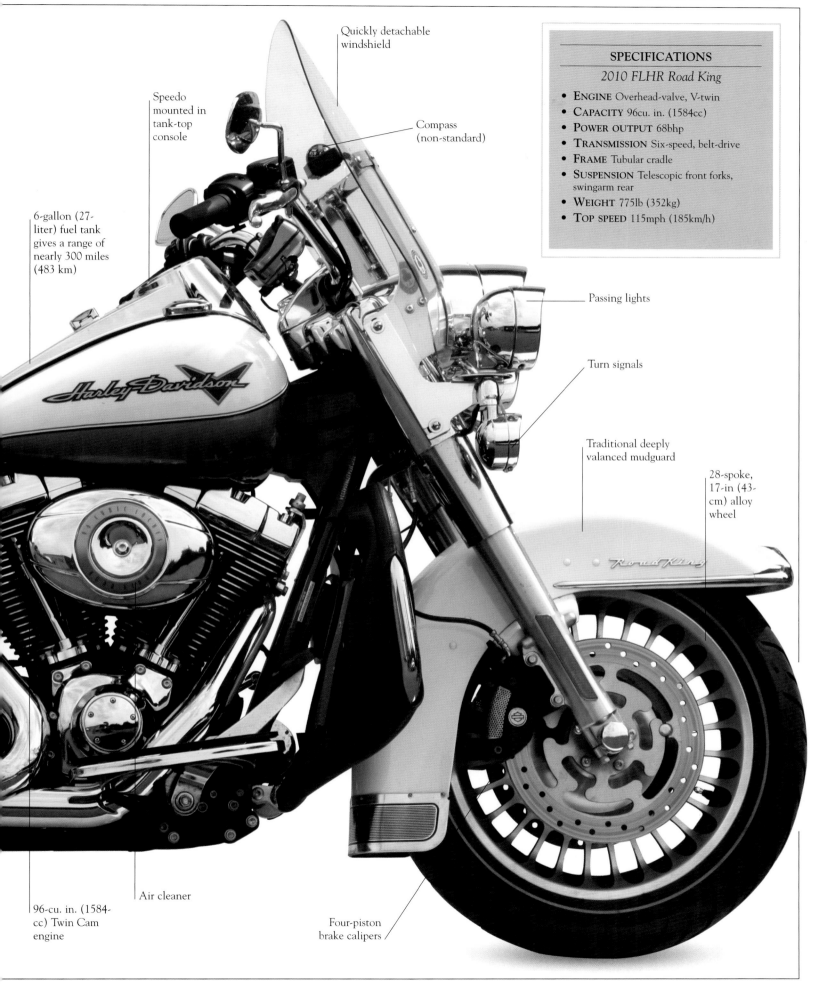

Quickly detachable
windshield

Speedo
mounted in
tank-top
console

Compass
(non-standard)

6-gallon (27-
liter) fuel tank
gives a range of
nearly 300 miles
(483 km)

Passing lights

Turn signals

Traditional deeply
valanced mudguard

28-spoke,
17-in (43-
cm) alloy
wheel

96-cu. in. (1584-
cc) Twin Cam
engine

Air cleaner

Four-piston
brake calipers

SPECIFICATIONS

2010 FLHR Road King

- **ENGINE** Overhead-valve, V-twin
- **CAPACITY** 96cu. in. (1584cc)
- **POWER OUTPUT** 68bhp
- **TRANSMISSION** Six-speed, belt-drive
- **FRAME** Tubular cradle
- **SUSPENSION** Telescopic front forks, swingarm rear
- **WEIGHT** 775lb (352kg)
- **TOP SPEED** 115mph (185km/h)

CHAPTER EIGHT

V-Rod

2001–2010

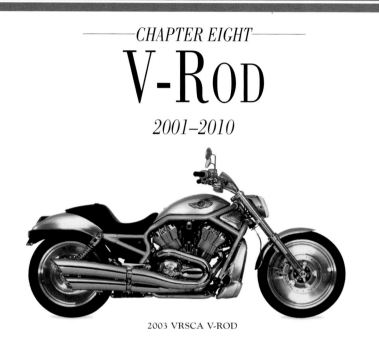

2003 VRSCA V-ROD

WHEN HARLEY-DAVIDSON introduced the V-Rod engine—officially called the "Revolution"—it came as something of a shock. Suddenly here was a Harley with overhead camshafts, water-cooling, and genuine performance. And the futuristic appearance of the V-Rod was also a surprise. Just when we thought that Harley had explored every variant of the cruiser look, the company invented something completely different.

A NEW DIRECTION
The radical nature of the V-Rod was the result of Harley responding to calls for significantly higher performance in an age of increasing noise and pollution restrictions.

2003 VRSCA V-Rod

HARLEY ENTHUSIASTS LOOKING for more performance from their bikes were finally rewarded in 2001 with the introduction of a radical new machine. The V-Rod was that rare thing in the history of the Milwaukee giant—a new model owing nothing to earlier machines. At the heart of the bike is an all-new 60° V-twin unit featuring overhead camshafts, four valves per cylinder, and water cooling, all innovations on a Harley engine. The chassis is long and low, with drag-bike-inspired looks and a unique finish in brushed and polished alloy. The V-Rod's astonishing performance and radical new look are exciting in themselves, but the promise they suggest for a new generation of Harley-Davidson motorcycles is remarkable.

SPECIFICATIONS
2003 VRSCA V-Rod

- **ENGINE** Dual overhead-cam, V-twin
- **CAPACITY** 1130cc
- **POWER OUTPUT** 115bhp @ 8,500rpm (claimed)
- **TRANSMISSION** Five-speed, belt drive
- **FRAME** Tubular steel perimeter
- **SUSPENSION** Telescopic front forks, swingarm rear
- **WEIGHT** 595lb (270kg) (claimed)
- **TOP SPEED** 130mph (209km/h)

2003 VRSCA V-Rod
With almost double the power of existing models, the V-Rod frame needed to be stiffer than any previous Harley frame. Large-diameter steel tubing is wrapped around the engine in what Harley calls a "perimeter" design, so that the frame is a styling component.

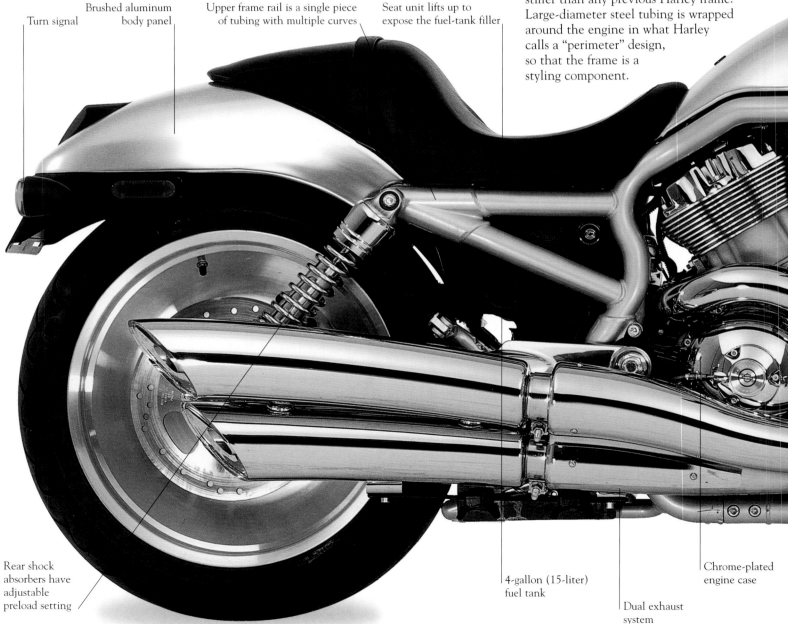

Turn signal

Brushed aluminum body panel

Upper frame rail is a single piece of tubing with multiple curves

Seat unit lifts up to expose the fuel-tank filler

Rear shock absorbers have adjustable preload setting

4-gallon (15-liter) fuel tank

Dual exhaust system

Chrome-plated engine case

Wing mirror

Instrument pod contains tachometer, speedometer, and fuel gauge

Dummy tank
The VRSCA's fuel tank cover is a dummy and actually conceals the airbox. The real fuel tank is positioned under the seat to lower the bike's center of gravity and contribute to its handling stability.

Distinctive oval headlight

49mm telescopic fork

Imposing rear view
The rear view of the V-Rod is dominated by the 180 section rear tire and the twin exhaust pipes, which both exit on the same side of the bike.

Solid disc aluminum wheel

Twin front disc brakes have 292mm discs and four-piston calipers

Cooling radiators are concealed in aluminum cowlings

Lower frame rail can be unbolted to aid engine removal

The V-Rod

TO MEET THE DEMAND FOR higher performance, Harley abandoned its traditional 45° overhead-valve design in favor of a 60° cylinder configuration in 2001. Consultants at Porsche worked with Harley engineers on the new V-Rod, with much of the technology originally developed for the VR1000 (*see pp. 176–77*). Liquid cooling, overhead valves, and twin camshafts in each head increased the power output. The "knife and fork" connecting rods, a feature of Harley V-twins since 1909, were replaced with side-by-side rods.

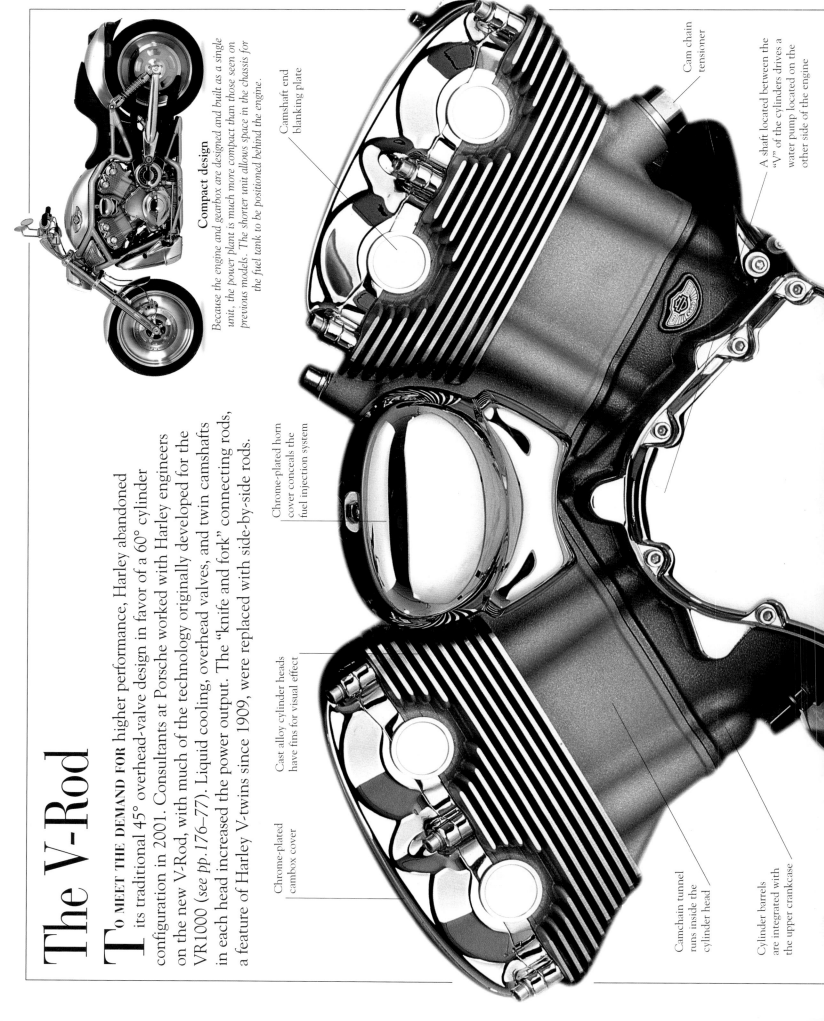

Compact design
Because the engine and gearbox are designed and built as a single unit, the power plant is much more compact than those seen on previous models. The shorter unit allows space in the chassis for the fuel tank to be positioned behind the engine.

Camshaft end blanking plate

Cam chain tensioner

A shaft located between the "V" of the cylinders drives a water pump located on the other side of the engine

Chrome-plated horn cover conceals the fuel injection system

Cast alloy cylinder heads have fins for visual effect

Chrome-plated cambox cover

Camchain tunnel runs inside the cylinder head

Cylinder barrels are integrated with the upper crankcase

Chrome-plated casing
conceals a compact
high output alternator

UP-TO-DATE TECHNOLOGY

The V-Rod uses modern technology in
manufacturing and design for optimum
performance within the constraints of
emissions and noise legislation, but also to
make the production process economic. Wet
sump lubrication, another unusual feature for a Harley,
means that oil is retained at the bottom of the engine,
so there is no need for a separate oil tank.

Camshaft bearing cap

Camshaft
drive sprocket

Camshaft

Valve

Cylinder head
gasket

Valve spring

Inlet port

Inside the cylinder head

*Twin camshafts run in the alloy cylinder head and operate two inlet
valves and two exhaust valves. The increased valve area compared
to the two-valve engines allows higher speeds and greater power.
Harley claim that the V-Rod produces 115bhp, far more than the
Evo (see pp.144–45) and Twin Cam (see pp.156–57) units.*

*"Up around 5,000rpm, where
your air-cooled hog is getting
flustered, the Rod takes on fresh
resolve. There's real urge to be
found from 6,000rpm on."*

BEN MILLER (BIKE MAGAZINE)

COMPETITION BIKE

• 2003 YAMAHA ROAD WARRIOR •
Yamaha was also developing powerful cruisers at
the time the V-Rod was introduced. The Road Star
Warrior has an old-fashioned OHV engine that relies
on huge capacity (1675cc) for its considerable power.

Water pump

Painted alloy
crankcases are
horizontally split

Fuel injection
air inlet

Clutch and
primary
drive gears

Engine mounting

Fuel injection system

*This cutaway drawing illustrates the V-Rod's fuel injection system,
with twin vertical air inlet tracts positioned between the cylinder
heads at the center of the "V."*

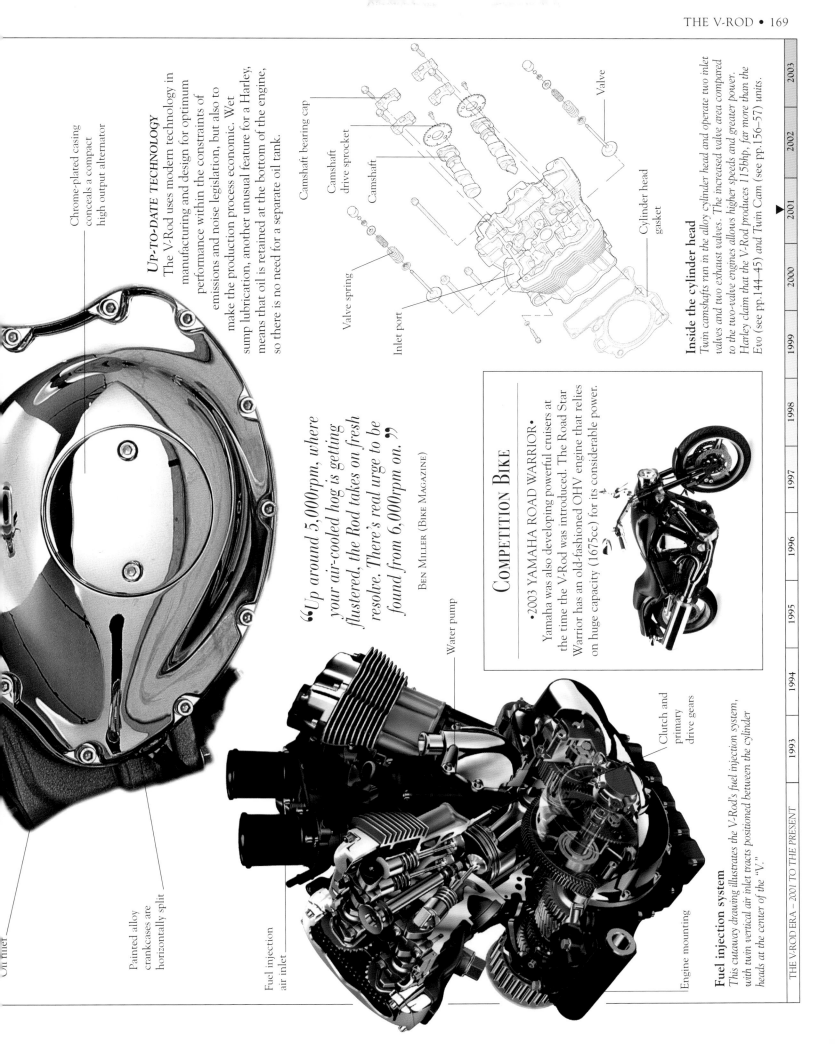

2010 CVO Softail Convertible

I**N 1999 THE MOTOR COMPANY** responded to the growing trend for customizing standard machines by creating their own limited-edition custom models. These were dubbed CVO models, for Custom Vehicle Operations. Every year a limited number of models were given the CVO treatment, with larger capacity engines than the base models, more expensive paint finishes, non-standard accessories, and performance parts from Harley's "Screamin' Eagle" catalog. The detailing reflected current trends in the custom world, with hand-applied pinstriping, metal-flake, and flame paint jobs being some of the finishes used. CVO models were naturally pricier than the standard bike, although fitting the parts at the factory was often more cost-effective than a retro fit.

CVO SOFTAIL CONVERTIBLE
Harley-Davidson's two traditional variants of the big-twin were the comfortable and well-equipped touring machines and the stylish cruisers. They came together in the Softail Convertible model which, with quickly detachable panniers and shield, could perform both roles: you could use it as a regular machine for short runs, or fit the shield and panniers, load up, and head out for longer rides.

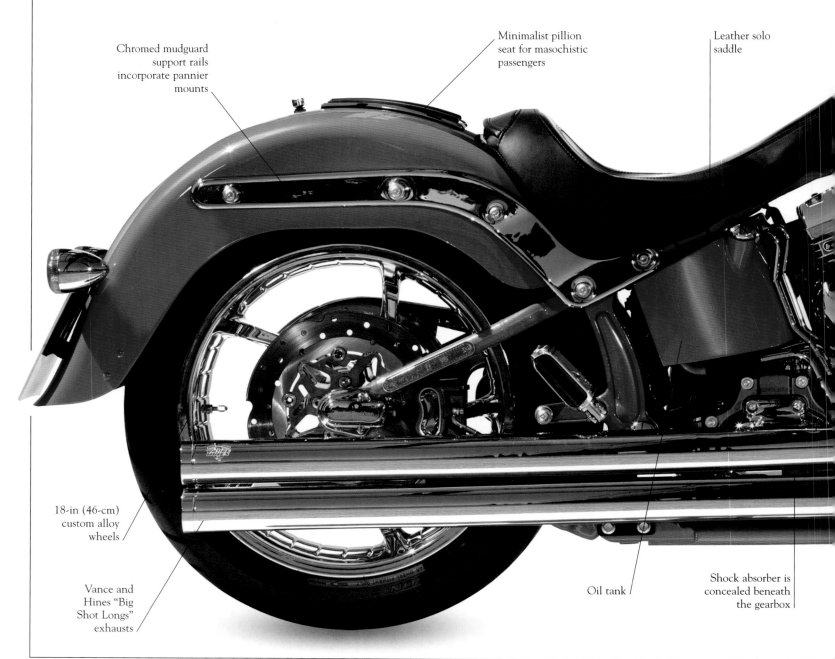

Chromed mudguard support rails incorporate pannier mounts

Minimalist pillion seat for masochistic passengers

Leather solo saddle

18-in (46-cm) custom alloy wheels

Vance and Hines "Big Shot Longs" exhausts

Oil tank

Shock absorber is concealed beneath the gearbox

5-gallon (19-liter) fuel tank with instrument console

Chromed handlebar controls

Small custom-style chrome headlight

Chromed fork shrouds echo 1950s Harley style

Deeply valanced FL-style mudguard

Single front disc brake with scalloped rotor

110-cu. in. (1804-cc) Twin Cam engine

"Screamin' Eagle" performance air filter

Billet alloy brake pedal

> **SPECIFICATIONS**
> *2010 CVO Softail Convertible*
> - **ENGINE** Overhead-valve, V-twin
> - **CAPACITY** 1804cc (110cu. in.)
> - **POWER OUTPUT** 85bhp
> - **TRANSMISSION** Six-speed, belt drive
> - **FRAME** Tubular cradle
> - **SUSPENSION** Telescopic front forks, swingarm rear
> - **WEIGHT** 724lb (328kg)
> - **TOP SPEED** 130mph (209km/h)

"Customizing is part of Harley-Davidson ownership—bikes don't remain the same as when they left the factory for long."

CHAPTER NINE

Sports Bikes

1994–2010

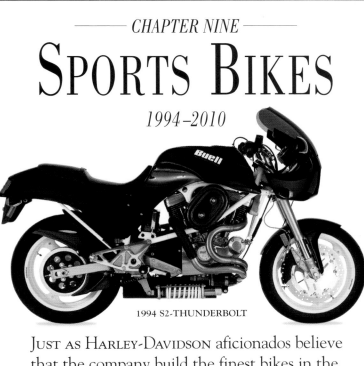

1994 S2-THUNDERBOLT

Just as Harley-Davidson aficionados believe that the company build the finest bikes in the world, those whose tastes don't run to traditional Harley-style cruisers dismiss them with as much vigor. In order to engage with the sports bike market, Harley bought a stake in a small company called Buell in 1993, which was making innovative Harley-engined machines. Harley also developed its own new 1000cc race bike. Sadly neither project ended well, with Buell production ending in 2009.

Hitting the Open Road

Harley-Davidson has traditionally been associated with touring bikes and cruisers, but owners of Harley-powered Buells were seeking performance and handling excellence too.

1986 Buell RR1000

THE FIRST **RR1000** PROTOTYPE was built in 1984 by Eric Buell, a former Harley employee, as a commission from the Vetter fairing company. Although Buell was still independent of Harley-Davidson at this point, the company would soon be incorporated into the Harley fold (*see pp.178–79*). The RR1000 used an XR1000 engine (*see pp.130–31*) mounted on Buell's patented Uniplanar chassis, which restricted engine vibration by using a system of rods, joints, and rubber mountings. Only 50 RR1000s were built before the supply of XR1000 engines dried up.

SPECIFICATIONS

1986 Buell RR1000

- **ENGINE** Overhead-valve, V-twin
- **CAPACITY** 61cu. in (998cc)
- **POWER OUTPUT** 77bhp @ 5,600rpm
- **TRANSMISSION** Four-speed, chain drive
- **FRAME** Chrome-moly, open-cradle space frame
- **SUSPENSION** Telescopic front forks, swingarm rear
- **WEIGHT** 390lb (177kg)
- **TOP SPEED** 135mph (217km/h) (est.)

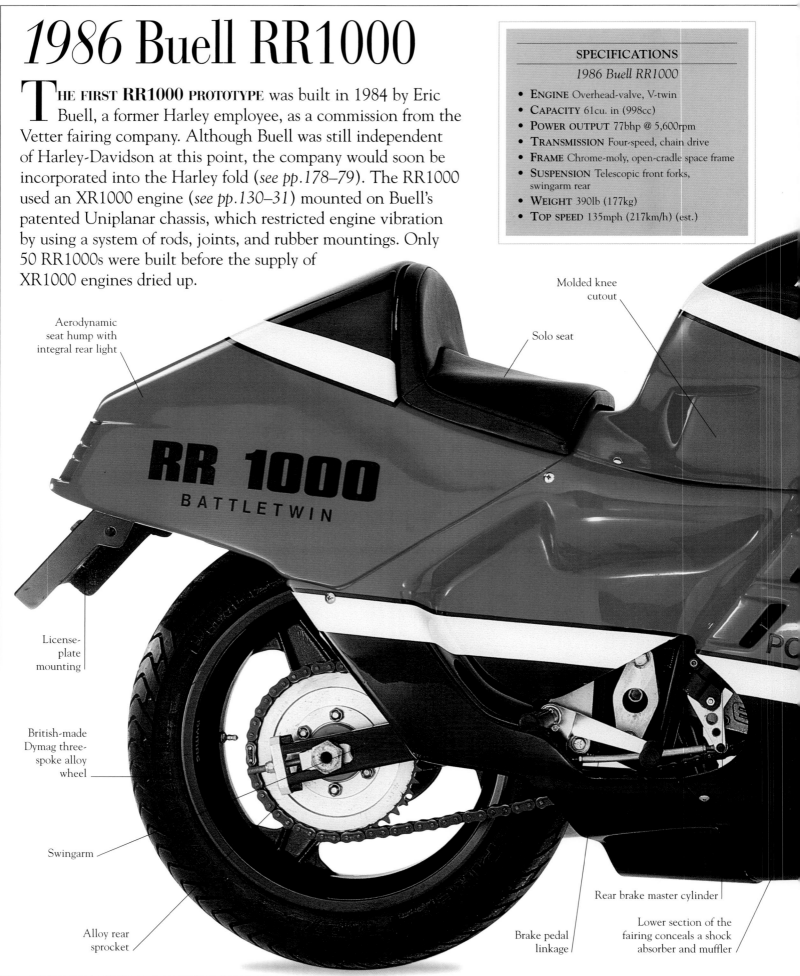

Molded knee cutout

Solo seat

Aerodynamic seat hump with integral rear light

License-plate mounting

British-made Dymag three-spoke alloy wheel

Swingarm

Alloy rear sprocket

Brake pedal linkage

Rear brake master cylinder

Lower section of the fairing conceals a shock absorber and muffler

Air link
tube for
forks

Fuel tank
breather pipe

Flush fuel
filler-cap

Aerodynamic
tinted
windshield

1986 BUELL RR1000 BATTLETWIN
Buell also offered a conventional 1200cc
Sportster engine version of the
Battletwin, which was designated the
RR1200. Though this larger-capacity
model was similar to the RR1000, the
exposed engine on the unfaired version—
the RS1200—resulted in a more
traditional looking machine.

Bar-end mirror

Harley-Davidson's
orange and black
racing color scheme

Oblong headlamp

Fairing mounting
bolt; turn signals
also positioned
here

Italian Marzocchi
telescopic forks

Aerodynamic
shrouded front
mudguard

Bulges reflect the position
of the carburetors on the
XR1000 engine

The Buell's Harley engine
has to be advertised,
otherwise you would never
know it was a Harley unit

Massive
front disc
brakes

16-in (41-cm)
wheel reflects
period fashion
for small wheels

Pirelli
MP7 tire

1994 VR1000

THERE ARE PROBABLY SEVERAL reasons why, in 1994, Harley decided to develop a totally new race bike. Corporate pride, the necessity to familiarize itself with new technology, and a desire to appeal to a new type of customer are among them. Whatever the reasons, the VR1000 made its debut under the spotlight at America's most prestigious road race—the 1994 Daytona 200-mile (322-km) Superbike. However, it wasn't a fairy-tale debut as the bike was off the pace and then blew up. Five years on, the VR1000 had still to achieve significant success despite swallowing large amounts of money and development time.

SPECIFICATIONS
1994 VR1000

- **ENGINE** Dual overhead-cam, V-twin
- **CAPACITY** 61cu. in. (996cc)
- **POWER OUTPUT** 140bhp @ 10,400rpm
- **TRANSMISSION** Six-speed, chain drive
- **FRAME** Twin-spar alloy
- **SUSPENSION** Inverted telescopic front fork, single-shock rear
- **WEIGHT** 355lb (161kg)
- **TOP SPEED** 190mph (306km/h) (Daytona gearing)

Lightweight alloy muffler

Molded racing seat

Tank cover and seat unit are a single lightweight structure

Braced swingarm

Slick racing tire provides maximum grip on dry tracks

Drilled rear disc brake

Underslung brake caliper

Alloy chassis is constructed with twin-spars connecting the steering head to the swingarm pivot

Two-into-one exhaust system

Exposed dry clutch is cable-operated

1994 VR1000

Limited numbers of the VR1000 were offered for sale to the public to comply with Superbike racing rules that stated that a number of production versions of the competing bikes had to be produced. This was the fourth bike off the production line and was ridden to third position by Ron McGill in the 1995 Bears Series.

Quick-detach fairing fasteners allow a replacement to be fitted during a race

Rider's gear
The Harley-Davidson "bar and shield" logo features heavily on Rider Ron McGill's leathers. Racing a VR1000 as a privateer was costly and most bikes weren't campaigned for a second season.

Quick-release strap for removing the tank cover and seat unit

Fairing designed by Willie G. Davidson

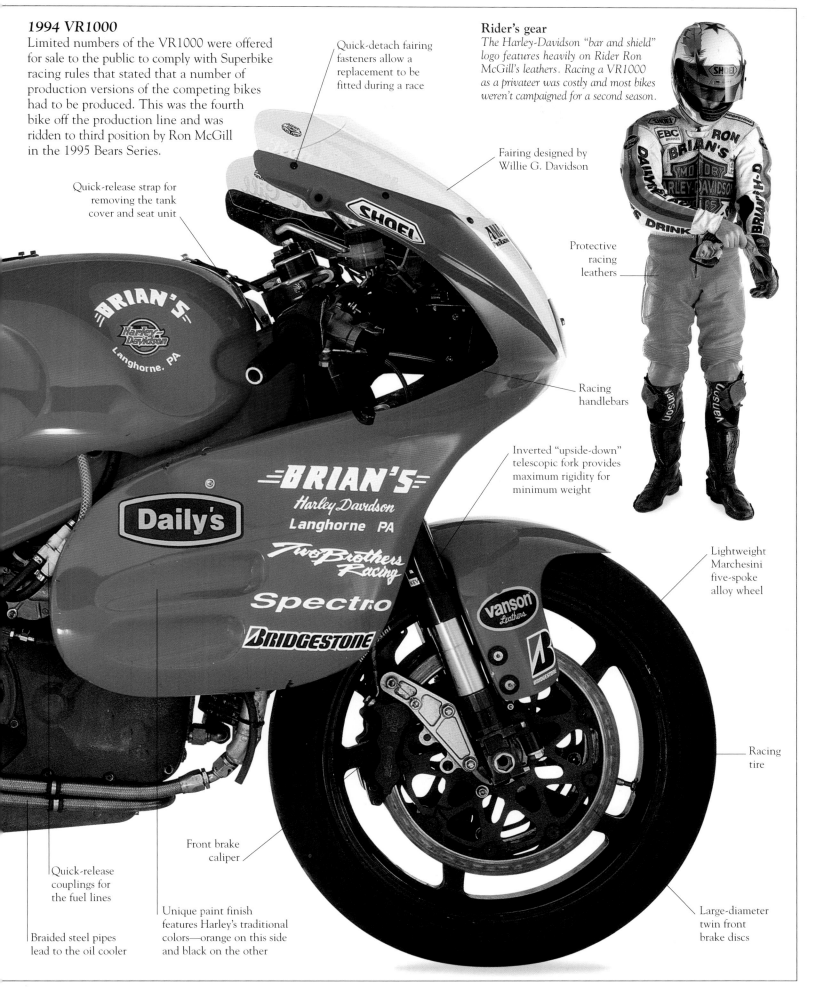

Protective racing leathers

Racing handlebars

Inverted "upside-down" telescopic fork provides maximum rigidity for minimum weight

Lightweight Marchesini five-spoke alloy wheel

Racing tire

Front brake caliper

Quick-release couplings for the fuel lines

Unique paint finish features Harley's traditional colors—orange on this side and black on the other

Large-diameter twin front brake discs

Braided steel pipes lead to the oil cooler

1994 Buell S2

I N 1993, HARLEY-DAVIDSON BOUGHT a 49 percent stake in US sports-bike manufacturer Buell, giving Harley the potential to explore new markets without alienating existing customers. Money was now available to develop new machines, and the S2-Thunderbolt was the first fruit of the new association. The bike was a development of the original Buell concept but with revised styling details and a 20 percent power increase thanks to improvements in the exhaust and intake systems. Production of Buell bikes increased from 100 to 700 per year, helping to make the Thunderbolt much cheaper than earlier Buells.

SPECIFICATIONS
1994 Buell S2

- ENGINE Overhead-valve, V-twin
- CAPACITY 73cu. in. (1203cc)
- POWER OUTPUT 76bhp
- TRANSMISSION Five-speed, belt drive
- FRAME Tubular cradle
- SUSPENSION Telescopic front forks, swingarm rear
- WEIGHT 450lb (204kg)
- TOP SPEED 110mph (177km/h)

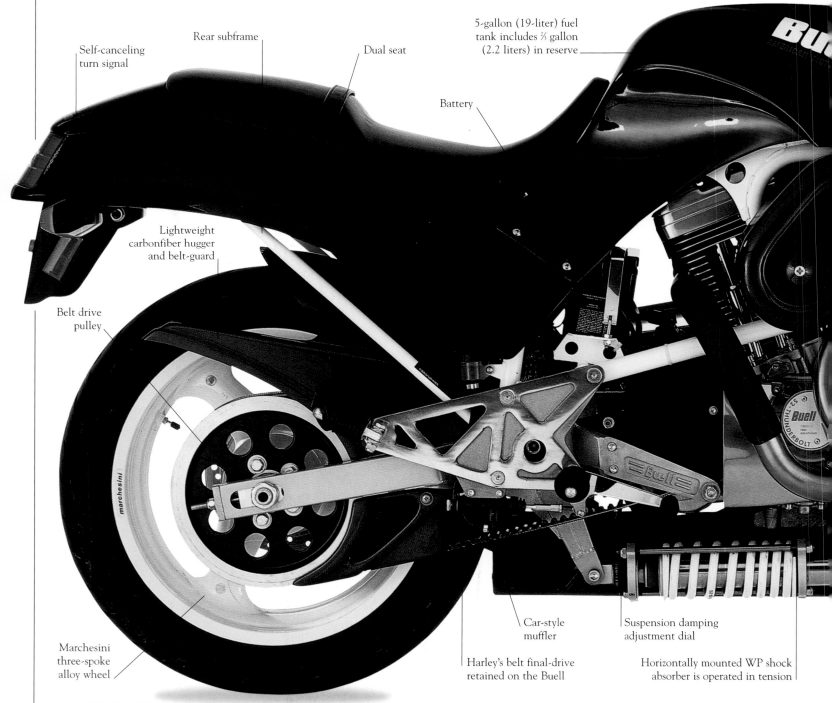

Self-canceling turn signal

Rear subframe

Dual seat

5-gallon (19-liter) fuel tank includes ⅗ gallon (2.2 liters) in reserve

Battery

Lightweight carbonfiber hugger and belt-guard

Belt drive pulley

Marchesini three-spoke alloy wheel

Car-style muffler

Harley's belt final-drive retained on the Buell

Suspension damping adjustment dial

Horizontally mounted WP shock absorber is operated in tension

1994 BUELL S2-THUNDERBOLT

Buell sourced high-quality components from outside suppliers for the Thunderbolt, including WP forks and shock absorbers, Marchesini wheels, and Performance Machine brake calipers. Combined with some well-thought-out engineering, the end result was a bike that handled very capably, even though the 1200cc Sportster engine was slightly underpowered for the Buell's chassis.

Aerodynamic tinted windshield

Chromed steel handlebars

Fuel breather pipe

Original Harley control lever

Instrument console

Frame-mounted bikini fairing

Headlight mounted in fairing

Rubber engine mounts reduce vibration

Short racing-style front mudguard

Two-into-one exhaust

Six-piston Performance Machine front brake caliper

13½-in (34-cm) front disc brake

Adjustable rearview mirror

Chrome-moly tubular frame

Air vents

Two-into-one stainless steel exhaust pipes

Dunlop Sportmax front tire

WP "upside-down" front forks

Slim profile
The Buell S2 made use of the Sportster power unit to provide the traditional benefits of the V-twin engine in a sport bike. This combination of a slim profile with good power delivery created a bike capable of competing with its contemporaries.

Knee cutout in fuel tank

Prop stand

Rear wheel is 1½in (3.8cm) wider than front wheel

1999 Buell X1 Lightning

WHEN **HARLEY BOUGHT ANOTHER** chunk of Buell—taking its stake in the sports bike company to 98 percent—in the late 1990s, the X1 Lightning followed soon afterward. Launched in 1998 for the 1999 model year, the Lightning was more polished and refined than earlier Buells, while retaining the oddball looks of the older machines. The ride was improved, the styling was cleaned up, and a new electronic fuel-injection system was added to the Sportster engine. But in refining the Lightning some of the raw charm of the earlier bikes was lost.

1999 BUELL X1 LIGHTNING

Showa suspension units front and rear gave the Lightning a far superior ride over previous models, and another neat touch was the alloy seat subframe. Mechanically, the new fuel-injection system increased outright power and allegedly made the delivery smoother. Harley-Davidson hoped that the X1 would attract a more mainstream buyer to its sports marque.

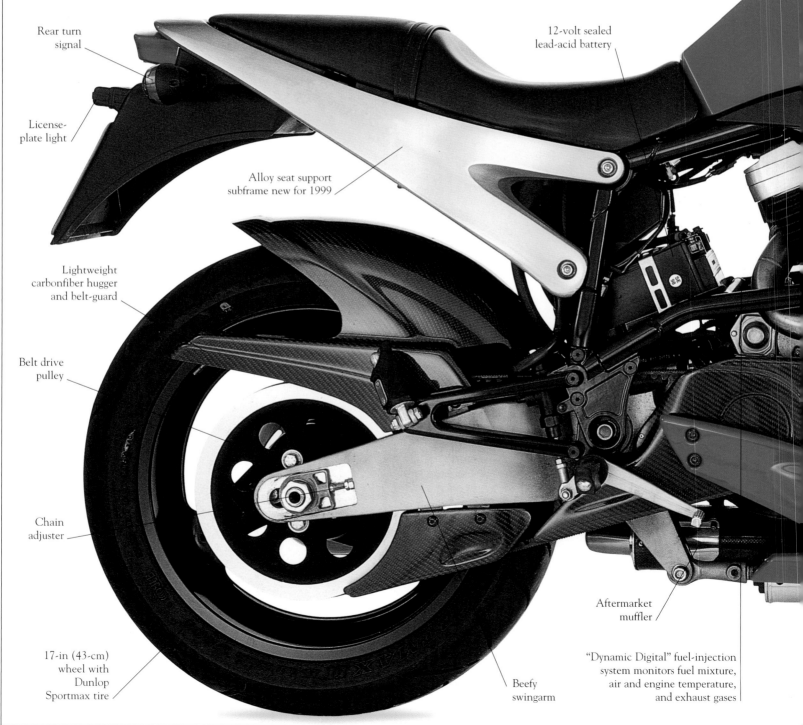

Rear turn signal

License-plate light

Alloy seat support subframe new for 1999

12-volt sealed lead-acid battery

Lightweight carbonfiber hugger and belt-guard

Belt drive pulley

Chain adjuster

17-in (43-cm) wheel with Dunlop Sportmax tire

Beefy swingarm

Aftermarket muffler

"Dynamic Digital" fuel-injection system monitors fuel mixture, air and engine temperature, and exhaust gases

Instruments include a speedometer, tachometer, odometer, and tripmeter

4⅗-gallon (17.4-liter) fuel tank

Fuel tank breather pipe

The design of the wing mirrors was new for 1999, allowing easier adjustment

"Chin" fairing

Turn signal

Chrome trimmed headlight

Nylon airbox

Six-piston brake calliper

Belly pan

Red Snap was one of four color schemes available; the others were Onyx Alloy, Reactor Yellow, and Carbon Black

Stainless steel two-into-one exhaust

"Upside down" Showa front forks are adjustable to suit different riders

Short racing-style front mudguard

Three-spoke cast-alloy wheel

SPECIFICATIONS
1999 Buell X1 Lightning

- **ENGINE** Overhead-valve, V-twin
- **CAPACITY** 73cu. in. (1203cc)
- **POWER OUTPUT** 101bhp @ 6,000rpm (claimed)
- **TRANSMISSION** Five-speed, belt drive
- **FRAME** Tubular cradle
- **SUSPENSION** Telescopic front forks, single-shock rear
- **WEIGHT** 441lb (200kg)
- **TOP SPEED** 140mph (225km/h)

" The Lightning incorporates one of the most advanced fuel management systems ever seen on a motorcycle. "

2002 Buell XB9R Firebolt

THE FIREBOLT WAS A radical new model introduced for 2002, and it marked Buell's first serious effort to produce a genuine sports bike. It was lighter, shorter, and more focused than their previous machines, with steeper steering geometry to provide the sharper steering response required by sports riders. To make the machine more compact, Buell's traditional tubular steel frame was replaced by a twin-spar alloy chassis. The Sportster-based engine had a reduced capacity, but a shorter stroke allowed the engine to rev to nearly 7,500rpm.

SPECIFICATIONS

2002 Buell XB9R Firebolt

- **ENGINE** Overhead-valve, V-twin
- **CAPACITY** 948cc
- **POWER OUTPUT** 92bhp @ 7,200rpm (claimed)
- **TRANSMISSION** Five-speed, belt drive
- **FRAME** Aluminum twin-spar
- **SUSPENSION** Telescopic front forks, swingarm rear
- **WEIGHT** 386lb (175kg) (dry, claimed)
- **TOP SPEED** 125mph (201km/h)

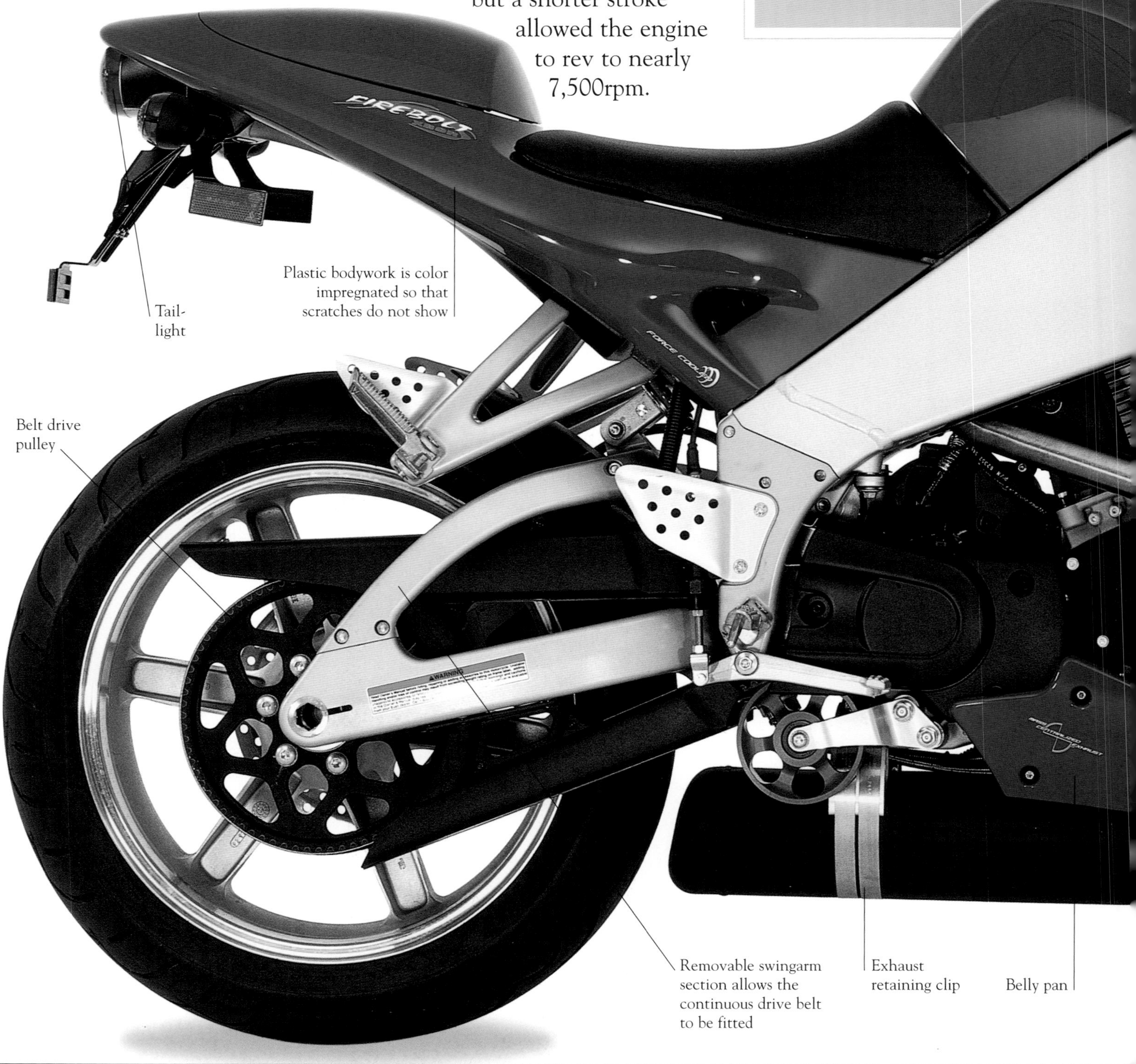

Tail-light

Plastic bodywork is color impregnated so that scratches do not show

Belt drive pulley

Removable swingarm section allows the continuous drive belt to be fitted

Exhaust retaining clip

Belly pan

2002 BUELL XB9R FIREBOLT

Buell designed the Firebolt around a new chassis and managed to give the bike a distinctive and good-looking profile. The new chassis wraps around the air-cooled engine, hiding the rear cylinder head from view. A fan helps keep the engine cool by drawing air past the rear cylinder. An oil cooler is mounted on the left side of the bike below the front cylinder.

Wing mirror

Clip-on handlebars are angled and low to provide a sporting riding position

"Upside down" Showa fork

Bikini fairing

3¾-gallon (14-liter) fuel tank

Distinctive appearance
Though sharing a slim frontal aspect, the Firebolt's bug-eye headlights and frame-mounted fairing differentiate it from Buell's Lightning model (see pp.184–85).

Dunlop tires developed for the bike are a special lightweight construction

"Inspired by the track. Built for the streets. The Firebolt comes armed with industry-leading innovations that are as ground-breaking as its style."

Rim-mounted disc brake is a unique Buell feature

2003 Buell XB9S Lightning

HAVING INTRODUCED THE radical new Firebolt for 2002 (*see pp.182–83*), Buell stripped the bike of its clip-on handlebars, frame-mounted sports fairing, and sports seat to create the aggressive-looking Lightning street-bike for the following year. The cast alloy rear sub-frame now supported a minimalist dual seat, and the higher bars and lower footrests provided a more accommodating riding position. Though these changes appear insignificant, and are relatively easy to implement, they alter the nature of the bike. The resulting model is less demanding and more practical than the uncompromising Firebolt it is based on. In terms of character and function, the Lightning has more in common with earlier Buells.

SPECIFICATIONS

2003 Buell XB9S Lightning

- **ENGINE** Overhead-valve, V-twin
- **CAPACITY** 984cc
- **POWER OUTPUT** 84bhp @ 7,400rpm (claimed)
- **TRANSMISSION** Five-speed, belt drive
- **FRAME** Aluminum twin-spar
- **SUSPENSION** Telescopic front forks, swingarm rear
- **WEIGHT** 386lb (175kg) (dry, claimed)
- **TOP SPEED** 120mph (193km/h)

2003 BUELL XB9S LIGHTNING

Particular care over choice of materials has been exercised in the production of the Lightning. The wheels use a combination of rough-cast and polished alloy for visual effect. Color-impregnated plastic body panels provide flashes of color, the engine uses a combination of black and bare alloy, and the exhaust pipe headers are made from stainless steel. The details are exquisite and passers-by find plenty to look at.

Low-profile dual seat

Cast alloy sub-frame supports the seat

Mudguard doubles as a license-plate mounting

❝ *The Lightning feels like a BMX, a bicycle with the last two pots of a Detroit V8 rumbling away instead of pedal power.* **❞**

17-in (43-cm) six-spoke rear wheel

7-in (180-mm) Dunlop tire

Drive belt tensioning pulley

Swingarm contains the oil tank (on the other side)

Plastic cover conceals air box

Wing mirror

Fork-mounted twin headlights

Tubular steel handlebars

Front section of the frame doubles as a fuel tank

Front forks are adjustable for preload and compression or rebound damping

Duct channels air into the engine

Oil cooler

Minimal protection
The central bulge of the fly-screen covers the tachometer, but the panel itself provides little in the way of weather protection for the rider.

Color options are yellow or black

Belly pan

Six-piston front brake caliper

2010 VRSCDX Night Rod Special

The water-cooled, four-valves per-cylinder and double overhead camshaft V-Rod engine offered a level of power and performance never seen on a production Harley-Davidson, but it also offered the potential for further tuning. With support from the factory, the bike became a popular choice for drag racing. The one make Harley Davidson racing series ran a V-Rod class, and the bike was extensively developed for the Pro-stock class, with the fastest machines achieving standing quarter mile times below seven seconds.

VRSCDX NIGHT ROD SPECIAL

The VRSCD Night Rod was introduced in 2006, its styling apparently inspired by Hot Rod cars and culture. The more powerful VRSCDX Night Rod Special, with an almost completely black finish, appeared the following year. The newer version had a 1250cc engine producing 125bhp. Power was transmitted to the road via a huge, 240-section rear tire, while uprated brakes meant efficient slowing too.

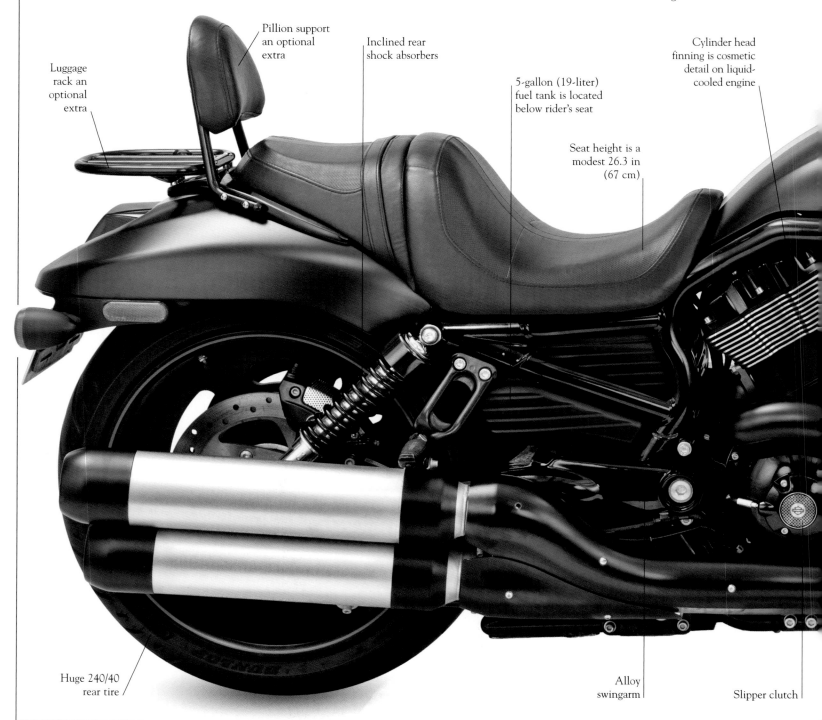

Pillion support an optional extra

Inclined rear shock absorbers

Cylinder head finning is cosmetic detail on liquid-cooled engine

Luggage rack an optional extra

5-gallon (19-liter) fuel tank is located below rider's seat

Seat height is a modest 26.3 in (67 cm)

Huge 240/40 rear tire

Alloy swingarm

Slipper clutch

Black finish on mirror housings and other components

Handlebar switches

Stainless steel "drag style" low rise handlebar

Dummy tank cover conceals airbox

Minimalist headlight fairing

SPECIFICATIONS
2010 VRSCDX Night Rod Special

- **ENGINE** Dual overhead cam, V-twin
- **CAPACITY** 76cu. in. (1250cc)
- **POWER OUTPUT** 125bhp
- **TRANSMISSION** Five-speed, belt drive
- **FRAME** Tubular steel perimeter
- **SUSPENSION** Telescopic front forks, swingarm rear
- **WEIGHT** 595lb (270kg)
- **TOP SPEED** 130mph (209km/h)

"The 125bhp VRSCDX is the fastest-accelerating production bike that Harley has ever built. So far at least."

Forks raked at 34 degrees

Front mudguard incorporating fork brace

300mm brake discs

19-in (48-cm) slotted alloy wheel

Water pump

Brake pedal

Four-piston Brembo brake caliper

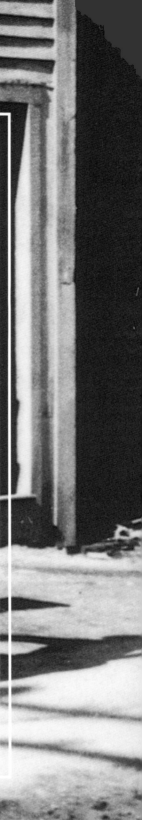

THE COMPLETE HARLEY-DAVIDSON
CATALOG
1903–2013

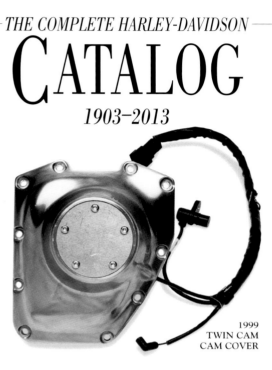

1999
TWIN CAM
CAM COVER

THE FOLLOWING CATALOG lists every production motorcycle manufactured by Harley-Davidson, from the first model in 1903 to the 2013 model range. It also includes various racing models produced by the company as one-offs or manufactured in limited production runs.

TEST DRIVING A NEW SILENT GRAY FELLOW
Harley-Davidson established a comprehensive dealership network only a few years after the Company was set up and today there are thousands of dealers worldwide.

The Complete Harley-Davidson Catalog

The following catalog lists the main models manufactured annually by Harley-Davidson from 1903 until the year 2013. It does not include any Buell models but does cover the Italian Aermacchi imports. Harley's philosophy during the 1980s of producing many different versions of similar models means that some limited editions and specialty models have been omitted. Note that the years given are model years so the bikes will actually have been released sometime during the previous year.

1903

The first year of Harley production, with an output of three bikes. The first bikes followed a 10.2cu. in.. (167cc) prototype that used a reinforced bicycle-type frame.

THE FIRST BIKE

- **ENGINE** F-HEAD SINGLE
- **CAPACITY** 24.74CU. IN..
- **NEW FEATURES** LOOP FRAME AND LARGER THAN USUAL ENGINE
- **COLORS** BLACK PAINT FINISH.

1904

First Harley-Davidson factory built in a 10-x-15-ft (3-x-4.5-m) shed and first dealer appointed.

MODEL 0

- **ENGINE** F-HEAD SINGLE
- **CAPACITY** 24.74CU. IN..
- **COLORS** BLACK PAINT FINISH. [START OF IDENTIFICATION NUMBERING SYSTEM THAT SUBTRACTS FOUR FROM THE YEAR TO GET THE MODEL NUMBER FOR THAT YEAR— E.G., 1904 = MODEL 0].

1905

Factory size doubled to 10 ft x 30 ft (3 m x 9 m). Walter Davidson becomes full-time factory manager.

MODEL 1

- **ENGINE** F-HEAD SINGLE
- **CAPACITY** 24.74CU. IN..
- **COLORS** BLACK PAINT FINISH.

1905 MODEL 1 SINGLE

1906

Silent Gray Fellow name first used for Harley's singles. New building on Juneau Avenue.

MODEL 2

- **ENGINE** F-HEAD SINGLE
- **CAPACITY** 26.8CU. IN..
- **NEW FEATURES** CAPACITY INCREASED TO 26.8CU. IN..
- **COLORS** BLACK PAINT FINISH (RENAULT GRAY OPTIONAL).

1907

First experimental Harley-Davidson V-twin is built and exhibited at a motorcycle show. Harley-Davidson Motor Company officially established.

MODEL 3

- **ENGINE** F-HEAD SINGLE
- **CAPACITY** 26.8CU. IN..
- **NEW FEATURES** "SAGER" SPRUNG FRONT FORK (USED TO 1948)
- **COLORS** RENAULT GRAY STANDARD (BLACK OPTIONAL).

1908

Walter Davidson wins the F.A.M. endurance run with a perfect score. The company now has 35 employees.

MODEL 4

- **ENGINE** F-HEAD SINGLE
- **CAPACITY** 26.8CU. IN..
- **COLORS** RENAULT GRAY STANDARD (BLACK OPTIONAL).

1909

V-twin cataloged for the first time, though the model disappeared from the range the following year.

MODEL 5 SINGLES

- **ENGINE** F-HEAD SINGLE
- **CAPACITY** 30.16CU. IN..

MODEL VARIATIONS

5-A, 5-B. 5-C

MODEL 5 V-TWIN

- **ENGINE** V-TWIN
- **CAPACITY** 49.48CU. IN..

MODEL VARIATIONS

5-D

- **NEW FEATURES** BOSCH MAGNETO IGNITION OFFERED SCHEBLER CARBURETOR OFFERED WIRE CONTROL CABLE USED
- **COLORS** RENAULT GRAY STANDARD (BLACK OPTIONAL).

1910

Belt idler added to models to give a form of clutch control. First year of a Harley racing model and acetylene lights as an option. The company now employs 149 workers.

MODEL 6

- **ENGINE** F-HEAD SINGLE
- **CAPACITY** 30.16CU. IN..
- **MODEL VARIATIONS** 6-A, 6-B. 6-C

RACING MODEL

6-E

- **NEW FEATURES** FIRST YEAR OF "BAR AND SHIELD" TRADEMARK LOGO
- **COLORS** RENAULT GRAY STANDARD (BLACK OPTIONAL).

1911

V-twin returns and remains continuously in the Harley range. Company expands to nearly 500 employees, producing over 5,500 bikes.

MODEL 7 SINGLES

- **ENGINE** F-HEAD SINGLE
- **CAPACITY** 30.16CU. IN..

MODEL VARIATIONS

7-A, 7-B, 7-C

MODEL 7 V-TWIN

- **ENGINE** V-TWIN
- **CAPACITY** 49.48CU. IN..

MODEL VARIATIONS

7-D

- **NEW FEATURES** MECHANICAL INLET VALVES ON V-TWIN
- **COLORS** RENAULT GRAY.

1912

All models have a new frame with the top tube sloping to the rear allowing a lower saddle which is mounted on a new sprung seat-post. Clutch assembly in rear hub on models with X designation. Workforce doubles.

MODEL 8 SINGLES

- **ENGINE** F-HEAD SINGLE
- **CAPACITY** 30.16CU. IN..

MODEL VARIATIONS

8-A, X-8, X-8A

1912 X-8 SILENT GRAY FELLOW

MODEL 8 V-TWINS

- **ENGINE** V-TWINS
- **CAPACITY** 49.48 & 61CU. IN..

MODEL VARIATIONS

8-D, X-8D (49.48CU. IN.. V-TWIN) X-8E (61CU. IN.. V-TWIN)

- **NEW FEATURES** CHAIN FINAL-DRIVE FIRST USED REAR HUB CLUTCH FIRST USED
- **COLORS** RENAULT GRAY STANDARD. FOUR OTHER COLOR OPTIONS ARE NOW OFFERED AT EXTRA COST.

1913

Harley-Davidson racing department founded. First year of Model G delivery van.

MODEL 9 SINGLES

- **ENGINE** F-HEAD SINGLE
- **CAPACITY** 35CU. IN..

MODEL VARIATIONS

9-A, 9-B

MODEL 9 V-TWINS

- **ENGINE** V-TWIN
- **CAPACITY** 61CU. IN..

MODEL VARIATIONS

9-E, 9-F, 9-G

- **NEW FEATURES** MECHANICAL INLET VALVES ON SINGLES MAGNETO IGNITION STANDARD TWO-SPEED TRANSMISSION OPTION AVAILABLE ON DELIVERY VAN

- **COLORS**
RENAULT GRAY STANDARD. FOUR COLOR OPTIONS AVAILABLE AT EXTRA COST.

1914

Harley-Davidson's factory race team starts competing on various circuits. First sidecars ordered from the Rogers Company, but not officially listed as an option until following year.

MODEL 10 SINGLES

- **ENGINE** F-HEAD SINGLE
- **CAPACITY** 35CU. IN..

MODEL VARIATIONS

10-A, 10-B, 10-C

MODEL 10 V-TWINS

- **ENGINE** V-TWIN
- **CAPACITY** 61CU. IN..

MODEL VARIATIONS

10-E, 10-F, 10-G

- **NEW FEATURES**
STEP (KICK) STARTERS OFFERED, ALLOWING THE BICYCLE PEDALS TO BE ABANDONED AND FOOTBOARDS FITTED TO SOME MACHINES
TWO-SPEED TRANSMISSION AVAILABLE FOR SINGLES

- **COLORS**
RENAULT GRAY STANDARD. FOUR COLOR OPTIONS AVAILABLE AT EXTRA COST.

1915

From 1915–53 a sidecar version of most of the V-twin models was offered. The initial S after the model ID letters was used to identify them— e.g., FS Sidecar Twin. Harley's new interest in bike racing spawns first series of racing models in both single and V-twin formats.

MODEL 11 SINGLES

- **ENGINE** F-HEAD SINGLE
- **CAPACITY** 35CU. IN..

MODEL VARIATIONS

11-B, 11-C

RACING MODELS

11-K4, 11-KF

MODEL 11 V-TWINS

- **ENGINE** V-TWIN
- **CAPACITY** 61CU. IN..

MODEL VARIATIONS

11-E, 11-F, 11-G, 11-H, 11-J

RACING MODELS

11-K, 11-KR, 11-KT, 11-KRH, 11-KTH, 11-K12, 11-K12H

- **NEW FEATURES**
THREE-SPEED GEARBOX ON TWINS
MECHANICAL OIL PUMP
ELECTRIC LIGHTING OPTION

- **COLORS**
RENAULT GRAY STANDARD. FOUR COLOR OPTIONS AVAILABLE AT EXTRA COST.

1915 KR FAST ROADSTER

1916

First year of eight-valve racing twins and four-valve single racers. Identification system changes this year so that the number reflects the model year, i.e. 1916 35cu. in.. single is a Model 16-B. [Note: Henceforth in this catalog, models will be referred to by identification initials only.]

MODEL B

- **ENGINE** F-HEAD SINGLE
- **CAPACITY** 35CU. IN..

MODEL C

- **ENGINE** F-HEAD SINGLE
- **CAPACITY** 35CU. IN..

RACING MODELS

MODEL S

MODEL E

- **ENGINE** F-HEAD V-TWIN
- **CAPACITY** 61CU. IN..

MODEL F

- **ENGINE** F-HEAD V-TWIN
- **CAPACITY** 61CU. IN..

MODEL J

- **ENGINE** F-HEAD V-TWIN
- **CAPACITY** 61CU. IN.

RACING MODELS

MODELS R & T
EIGHT-VALVE RACER (OHV)

- **COLORS** RENAULT GRAY STANDARD. FOUR COLOR OPTIONS AVAILABLE AT EXTRA COST.

1917

Bosch magnetos unavailable due to the war, so replaced by Dixie brand for this year only. Harley sends its first bikes for military use in WWI.

MODEL B

- **ENGINE** F-HEAD SINGLE
- **CAPACITY** 35CU. IN.

MODEL C

- **ENGINE** F-HEAD SINGLE
- **CAPACITY** 35CU. IN.

RACING MODELS

MODEL S

MODEL E

- **ENGINE** F-HEAD V-TWIN
- **CAPACITY** 61CU. IN.

MODEL F

- **ENGINE** F-HEAD V-TWIN
- **CAPACITY** 61CU. IN.

MODEL J

- **ENGINE** F-HEAD V-TWIN
- **CAPACITY** 61CU. IN.

RACING MODELS

MODELS R & T, EIGHT-VALVE RACER

- **COLORS**
OLIVE GREEN IS NOW THE STANDARD COLOR. ONLY YEAR OF GOLD PINSTRIPING.

1918

Last year of the 35cu. in. F-head singles. Berling magnetos replace Dixies. Model F is the best-seller for this year.

MODEL B

- **ENGINE** F-HEAD SINGLE
- **CAPACITY** 35CU. IN.

MODEL C

- **ENGINE** F-HEAD SINGLE
- **CAPACITY** 35CU. IN.

MODEL E

- **ENGINE** F-HEAD V-TWIN
- **CAPACITY** 61CU. IN.

MODEL F

- **ENGINE** F-HEAD V-TWIN
- **CAPACITY** 61CU. IN.

MODEL J

- **ENGINE** F-HEAD V-TWIN
- **CAPACITY** 61CU. IN.

1918 MODEL J SIDECAR

MODEL VARIATIONS

FUS (GOVT. USE ONLY)

RACING MODELS

MODEL R, EIGHT-VALVE RACER

A SMALL NUMBER OF MODIFIED V-TWINS ALSO PRODUCED

- **COLORS**
OLIVE GREEN.

1919

First year of the horizontally opposed flathead twin model W, though not a great success in the US so mainly exported to overseas markets. Also first year for the "two-cam" F-head and eight-valve racers.

MODEL W

- **ENGINE** HORIZONTALLY OPPOSED FLATHEAD TWIN
- **CAPACITY** 35.64CU. IN. (584.03cc)

1919 MODEL W

MODEL F

- **ENGINE** F-HEAD V-TWIN
- **CAPACITY** 61CU. IN.

MODEL VARIATIONS

F/FS/FUS (GOVT. USE ONLY)

MODEL J

- **ENGINE** F-HEAD V-TWIN
- **CAPACITY** 61CU. IN.

MODEL VARIATIONS

J/JS

RACING MODELS

TWO-CAM EIGHT-VALVE RACER
TWO-CAM F-HEAD RACER

A SMALL NUMBER OF ADAPTED V-TWIN MODELS ALSO PRODUCED

- **COLORS**
OLIVE GREEN.

1920

Harley factory race team parades on victory lap with a pig on a bike, so beginning the "hog" association. The company now has the biggest motorcycle factory in the world.

MODEL W

- **ENGINE** HORIZONTALLY OPPOSED FLATHEAD TWIN
- **CAPACITY** 35.64CU. IN. (584.03cc)

MODEL VARIATIONS

W/WF/WJ

MODEL F

- **ENGINE** F-HEAD V-TWIN
- **CAPACITY** 61CU. IN.

MODEL VARIATIONS

F/FS

MODEL J

- **ENGINE** F-HEAD V-TWIN
- **CAPACITY** 61CU. IN.

MODEL VARIATIONS

J/JS

RACING MODELS

Two-cam eight-valve racer
Two-cam F-head racer

1920 EIGHT-VALVE RACER

A SMALL NUMBER OF MODIFIED
V-TWINS ALSO PRODUCED

- COLORS
OLIVE GREEN.

1921

First year for 37cu. in. CD
commercial single model, using
same chassis as (also new) 74cu. in.
big-twin but minus one cylinder.
A Harley racer is the first bike to
win a race with an average speed
of over 100mph (161km/h).

MODEL CD

- ENGINE F-HEAD SINGLE

- CAPACITY 37CU. IN.

MODEL W

- ENGINE HORIZONTALLY OPPOSED
FLATHEAD TWIN

- CAPACITY 35.64CU. IN. (584.03CC)

MODEL VARIATIONS

W/WF/WJ

MODEL F

- ENGINE F-HEAD V-TWIN

- CAPACITY 61CU. IN.

MODEL VARIATIONS

F/FS

MODEL J

- ENGINE F-HEAD V-TWIN

- CAPACITY 61CU. IN.

MODEL VARIATIONS

J/JS

MODEL FD

- ENGINE F-HEAD V-TWIN

- CAPACITY 74CU. IN.

MODEL VARIATIONS

FD/FDS

MODEL JD

- ENGINE F-HEAD V-TWIN

- CAPACITY 74CU. IN.

MODEL VARIATIONS

JD/JDS (SIDECAR TWINS)

RACING MODELS

Two-cam eight-valve racer
Two-cam F-head racer
Four-valve racer

A SMALL NUMBER OF MODIFIED
V-TWINS ALSO PRODUCED

- COLORS
OLIVE GREEN. ONLY YEAR OF OLIVE
GREEN CRANKCASES.

1922

Model CD commercial single
discontinued at the end of this year.
Harley-Davidson output picks up
again after national economic slump
of 1921 which saw sales drop by
60 percent and many other
motorcycle manufacturers go
out of business.

MODEL CD

- ENGINE F-HEAD SINGLE

- CAPACITY 37CU. IN.

MODEL W

- ENGINE HORIZONTALLY OPPOSED
FLATHEAD TWIN

- CAPACITY 35.64CU. IN.
(584.03CC)

MODEL VARIATIONS

W/WF/WJ

MODEL F

- ENGINE F-HEAD V-TWIN

- CAPACITY 61CU. IN.

MODEL VARIATIONS

J/JS

MODEL J

- ENGINE F-HEAD V-TWIN

- CAPACITY 61CU. IN.

MODEL VARIATIONS

F/FS

MODEL FD

- ENGINE F-HEAD V-TWIN

- CAPACITY 74CU. IN.

- MODEL VARIATIONS

FD/FDS

MODEL JD

- ENGINE F-HEAD V-TWIN

- CAPACITY 74CU. IN.

- MODEL VARIATIONS

JD/JDS

RACING MODELS

Two-cam eight-valve racer
Two-cam F-head racer
Four-valve racer

A SMALL NUMBER OF MODIFIED
V-TWINS ALSO PRODUCED

- COLORS
BREWSTER GREEN PAINT SCHEME
(INCLUDING CRANKCASES).

1923

Last year of W Series sport twins, with
production figures only in the
hundreds. Harley factory race team is
disbanded. JD model is best-seller,
with figures of over 7,000. First year of
hinged rear mudguard on big-twins.

MODEL W

- ENGINE HORIZONTALLY OPPOSED
FLATHEAD TWIN

- CAPACITY 35.64CU. IN. (584.03CC)

MODEL VARIATIONS

W/WF/WJ

MODEL F

- ENGINE F-HEAD V-TWIN

- CAPACITY 61CU. IN.

MODEL VARIATIONS

F/FS

MODEL J

- ENGINE F-HEAD V-TWIN

- CAPACITY 61CU. IN.

MODEL VARIATIONS

J/JS

MODEL FD

- ENGINE F-HEAD V-TWIN

- CAPACITY 74CU. IN.

MODEL VARIATIONS

FD/FDS

MODEL JD

- ENGINE F-HEAD V-TWIN

- CAPACITY 74CU. IN.

MODEL VARIATIONS

JD, JDS

RACING MODELS

MODEL T (V-TWIN RACER)
MODEL S (SINGLE RACER)
TWO-CAM EIGHT-VALVE RACER
TWO-CAM F-HEAD RACER

A SMALL NUMBER OF MODIFIED
V-TWINS ALSO PRODUCED

- COLORS
BREWSTER GREEN PAINT SCHEME.

1924

Generally a poor year for the
company, which operates at a loss.
As a result, 1,500 employees are
laid off. Aluminum alloy pistons
used on some models for
this year only.

MODEL F

- ENGINE F-HEAD V-TWIN

- CAPACITY 61CU. IN.

MODEL VARIATIONS

F/FE/FES

MODEL J

- ENGINE F-HEAD V-TWIN

- CAPACITY 61CU. IN.

MODEL VARIATIONS

J/JE/JES

MODEL FD

- ENGINE F-HEAD V-TWIN

- CAPACITY 74CU. IN.

MODEL VARIATIONS

FD/FDS/FDCA/FDSCA

MODEL JD

- ENGINE F-HEAD V-TWIN

- CAPACITY 74CU. IN.

MODEL VARIATIONS

JD/JDS/JDCA/JDSCA

RACING MODELS

TWO-CAM EIGHT-VALVE RACER
TWO-CAM F-HEAD RACER

- COLORS
OLIVE GREEN.

1925

Last year that Harley outsources
sidecar manufacture, the company
now deciding to build its own. First
export of Harleys to Japan. Revamp of
the big-twins, with major mechanical
and design changes including Harley's
famous teardrop fuel tank. Despite the
decline in popularity of board-track
racing, Joe Petrali secures a number of
notable victories on a Harley.

MODEL F

- ENGINE F-HEAD V-TWIN

- CAPACITY 61CU. IN.

MODEL VARIATIONS

F/FE/FES

MODEL J

- ENGINE F-HEAD V-TWIN

- CAPACITY 61CU. IN.

MODEL VARIATIONS

J/JE/JES

MODEL FD

- ENGINE F-HEAD V-TWIN

- CAPACITY 74CU. IN.

MODEL VARIATIONS

FD/FDCB/FDSBS

MODEL JD

- ENGINE F-HEAD V-TWIN

- CAPACITY 74CU. IN.

MODEL VARIATIONS

JD/JDCB/JDCBS

RACING MODELS

TWO-CAM EIGHT-VALVE RACER
TWO-CAM F-HEAD RACER

• NEW FEATURES
WIDER, LOWER FRAME
CYLINDRICAL TOOLBOX MOUNTED
ON FRONT FORK TEARDROP-STYLE
STREAMLINED TANKS
IRON ALLOY PISTONS

• COLORS
OLIVE GREEN.

1926

Models A and B are Harley's new flathead singles, with optional overhead-valve units on models AA and BA. All roadgoing twins now equipped with electrical lighting. First year of Harley's own-production sidecars. Over 9,000 JD models manufactured this year, making it the most popular Harley by far.

MODEL A

• ENGINE FLATHEAD SINGLE

• CAPACITY 21CU. IN.

MODEL VARIATIONS

A, AA (OHV ENGINE)

MODEL B

• ENGINE FLATHEAD SINGLE

• CAPACITY 21CU. IN.

MODEL VARIATIONS

B, BA (OHV ENGINE)

1926 B PEASHOOTER

MODEL F

• ENGINE F-HEAD V-TWIN

• CAPACITY 61CU. IN.

MODEL VARIATIONS

F/FE/FES

MODEL J

• ENGINE F-HEAD V-TWIN

• CAPACITY 61CU. IN.

MODEL VARIATIONS

J/JE/JES/JS

MODEL FD

• ENGINE F-HEAD V-TWIN

• CAPACITY 74CU. IN.

MODEL VARIATIONS

FD/FDCB/FDSBS

MODEL JD

• ENGINE F-HEAD V-TWIN

• CAPACITY 74CU. IN.

MODEL VARIATIONS

JD/JDCB/JDCBS/JDS

RACING MODELS

MODEL S (21 CU. IN. OHV SINGLE)
TWO-CAM F-HEAD RACER
TWO-CAM EIGHT-VALVE RACER

1926 MODEL S RACER

• COLORS
OLIVE GREEN. WHITE OR CREAM
AVAILABLE AT EXTRA COST ON TWINS.

1927

A year of minimal change, with only minor detail modifications, though models AA and BA do get Ricardo cylinder heads and a revised frame.

MODEL A

• ENGINE FLATHEAD SINGLE

• CAPACITY 21CU. IN.

MODEL VARIATIONS

A; AA/AAE (OHV ENGINE)

MODEL B

• ENGINE FLATHEAD SINGLE

• CAPACITY 21CU. IN.

MODEL VARIATIONS

B; BA/BAE (OHV ENGINE)

MODEL J

• ENGINE F-HEAD V-TWIN

• CAPACITY 61CU. IN.

MODEL VARIATIONS

J/JS

MODEL F

• ENGINE F-HEAD V-TWIN

• CAPACITY 61CU. IN.

MODEL VARIATIONS

F/FK/FS

MODEL JD

• ENGINE F-HEAD V-TWIN

• CAPACITY 74CU. IN.

MODEL VARIATIONS

JD/JDL/JDS

MODEL FD

• ENGINE F-HEAD V-TWIN

• CAPACITY 74CU. IN.

MODEL VARIATIONS

FD/FDL/FDS

RACING MODELS

MODELS S/SM/SA/SMA (21 CU. IN.
OHV SINGLES)
MODEL T (V-TWIN)
MODELS FHAC/FHAD (61 CU. IN.
HILL-CLIMBERS)

TWO-CAM EIGHT-VALVE RACER
TWO-CAM F-HEAD RACER

• COLORS
OLIVE GREEN. WHITE OR CREAM
AVAILABLE AT EXTRA COST ON TWINS.

1928

New high-performance JH and JDH models are roadgoing versions of the two-cam competition machines, using a slimmer and lower chassis than the standard single-cam J and JD models.

MODEL A

• ENGINE FLATHEAD SINGLE

• CAPACITY 21CU. IN.

MODEL VARIATIONS

A; AA/AAE (OHV ENGINE)

MODEL B

• ENGINE FLATHEAD SINGLE

• CAPACITY 21CU. IN.

MODEL VARIATIONS

B; BA/BAE (OHV ENGINE)

MODEL J

• ENGINE F-HEAD V-TWIN

• CAPACITY 61CU. IN.

MODEL VARIATIONS

J/JS/JX/JXL/JH

MODEL F

• ENGINE F-HEAD V-TWIN

• CAPACITY 61CU. IN.

MODEL VARIATIONS

F/FH/FS

MODEL JD

• ENGINE F-HEAD V-TWIN

• CAPACITY 74CU. IN.

MODEL VARIATIONS

JD/JDH (TWO-CAM)/JDX/JDXL/JDS

1928 JD

MODEL FD

• ENGINE F-HEAD V-TWIN

• CAPACITY 74CU. IN.

MODEL VARIATIONS

FD/FDH/FDS

RACING MODELS

MODELS S/SM/SA/SMA (21 CU. IN.
OHV SINGLES)
MODEL T (V-TWIN)
MODEL FHAC/FHAD

(61 CU. IN. HILL-CLIMBERS)
TWO-CAM EIGHT-VALVE RACER
TWO-CAM F-HEAD RACER

• NEW FEATURES
FRONT BRAKE

• COLORS
OLIVE GREEN. A RANGE OF EIGHT
COLORS AVAILABLE AT EXTRA COST
ON TWINS.

1929

New model C single features a bigger engine. Also first year of model D 45 cu. in. side-valve V-twin. Last year of AA and BA singles, models J, JD, F, and FD, as well as eight-valve racers.

MODEL A

• ENGINE FLATHEAD SINGLE

• CAPACITY 21CU. IN.

MODEL VARIATIONS

A; AA (OHV ENGINE)

MODEL B

• ENGINE FLATHEAD SINGLE

• CAPACITY 21CU. IN.

MODEL VARIATIONS

B/BA; BAF (OHV ENGINE)

MODEL C

• ENGINE FLATHEAD SINGLE

• CAPACITY 30.5CU. IN.

MODEL D

• ENGINE FLATHEAD V-TWIN

• CAPACITY 45CU. IN.

MODEL VARIATIONS

D/DL

MODEL J

• ENGINE F-HEAD V-TWIN

• CAPACITY 61CU. IN.

MODEL VARIATIONS

J/JS/JXL/JH

MODEL F

• ENGINE F-HEAD V-TWIN

• CAPACITY 61CU. IN.

MODEL VARIATIONS

F/FL/FS

MODEL JD

• ENGINE F-HEAD V-TWIN

• CAPACITY 74CU. IN.

MODEL VARIATIONS

JD/JDH(TWO-CAM)//JDF/JDXL/JDS

MODEL FD

• ENGINE F-HEAD V-TWIN

• CAPACITY 74CU. IN.

MODEL VARIATIONS

FD/FDH/FDL/FDS

RACING MODELS

SA/SMA (21cu. in. ohv singles)
Model T (V-twin)
FHAC (61cu. in. racer)
FHAD (61cu. in. hill-climber)
Two-cam eight-valve racer
Two-cam F-head racer

• NEW FEATURES
DUAL BULLET HEADLIGHTS
FOUR-TUBE "PIPE O' PAN" SILENCERS

• COLORS
OLIVE GREEN. A RANGE OF EIGHT
COLORS AVAILABLE AT EXTRA COST
ON TWINS.

1930

New flathead model V replaces F-head big-twins, boasting magnesium-alloy pistons and Ricardo heads.

MODEL A

• ENGINE FLATHEAD SINGLE

• CAPACITY 21cu. in.

MODEL B

• ENGINE FLATHEAD SINGLE

• CAPACITY 21cu. in.

MODEL VARIATIONS

B/BR; BAF (OHV ENGINE)

MODEL C

• ENGINE FLATHEAD SINGLE

• CAPACITY 30.5cu. in.

MODEL VARIATIONS

C/CM

MODEL D

• ENGINE FLATHEAD V-twin

• CAPACITY 45cu. in.

• MODEL VARIATIONS

D/DL/DLD/DS

MODEL V

• ENGINE FLATHEAD V-twin

• CAPACITY 74cu. in.

MODEL VARIATIONS

V/VS/VL/VLM/VC/VCM/VM/
VMS/VMG

RACING MODELS

HILL-CLIMBER AND OTHER RACERS
WERE MADE AS ONE-OFF BIKES
THROUGHOUT THE 1930s

1930 HILL-CLIMBER

• NEW FEATURES
DROP-CENTER WHEEL RIMS
FORGED "I" BEAM FORKS

• COLORS
OLIVE GREEN. A RANGE OF SIX
COLORS AVAILABLE AT EXTRA COST
ON TWINS.

1931

Model B produced for export only this year. Last year of model D. Low production levels in post-crash slump.

MODEL B

• ENGINE FLATHEAD SINGLE

• CAPACITY 21cu. in.

MODEL C

• ENGINE FLATHEAD SINGLE

• CAPACITY 30.5cu. in.

MODEL VARIATIONS

C/CH/CMG/CC

MODEL D

• ENGINE FLATHEAD V-twin

• CAPACITY 45cu. in.

MODEL VARIATIONS

D/DL/DLD/DC/DS

MODEL V

• ENGINE FLATHEAD V-twin

• CAPACITY 74cu. in.

MODEL VARIATIONS

VVS/VL/VLM/VC/VSR/VMG/VMS

• NEW FEATURES
"SUNBURST" HORN (ALL MODELS
EXCEPT B)
SCHEBLER DELUXE CARBURETOR
ON TWINS

• COLORS
OLIVE GREEN. A RANGE OF FIVE
COLORS AVAILABLE AT EXTRA COST.

1932

New model R replaces model D. First year of three-wheeled model G Servi-Car. Model B produced for domestic market again.

MODEL B

• ENGINE FLATHEAD SINGLE

• CAPACITY 21cu. in.

MODEL C

• ENGINE FLATHEAD SINGLE

• CAPACITY 30.5cu. in.

MODEL VARIATIONS

C/CC/CR/CS

MODEL R

• ENGINE FLATHEAD V-twin

• CAPACITY 45cu. in.

MODEL VARIATIONS

R/RL/RLD/RS

MODEL G SERVI-CAR

• ENGINE FLATHEAD V-twin

• CAPACITY 45cu. in.

MODEL VARIATIONS

G/GA/GD/GE

MODEL V

• ENGINE FLATHEAD V-twin

• CAPACITY 74cu. in.

MODEL VARIATIONS

V/VS/VL/VC

• COLORS
OLIVE GREEN. A RANGE OF SEVEN
COLORS AVAILABLE AT EXTRA COST.

1933

Layoffs and reduced production in 1932 mean little change in 1933. Joe Petrali adds a note of optimism to the gloom by winning the second of four consecutive national hill-climbing championships on a Harley.
He won six in total.

MODEL B

• ENGINE FLATHEAD SINGLE

• CAPACITY 21cu. in.

MODEL C

• ENGINE FLATHEAD SINGLE

• CAPACITY 30.5cu. in.

MODEL VARIATIONS

C/CB/CS

MODEL R

• ENGINE FLATHEAD V-twin

• CAPACITY 45cu. in.

MODEL VARIATIONS

R/RL/RLD/RS/RE/RLE/RLDE

MODEL G SERVI-CAR

• ENGINE FLATHEAD V-twin

• CAPACITY 45cu. in.

MODEL VARIATIONS

G/GA/GD/GDT/GE

MODEL V

• ENGINE FLATHEAD V-twin

• CAPACITY 74cu. in.

MODEL VARIATIONS

V/VS/VL/VLD/VC/VE/VF/VFS/
VLE/VSE

1933 VLE

• NEW FEATURES
ART DECO-STYLE BIRD MOTIF ON
TANK (THIS YEAR ONLY)
REVERSE GEAR OPTIONAL ON
R MODELS
BUDDY SEAT DEBUTS AS ACCESSORY

• COLORS OLIVE GREEN NO
LONGER STANDARD COLOR. SINGLES
AND MODEL G IN SILVER AND
TURQUOISE. TWINS IN A RANGE OF
FIVE STANDARD COLORS.

1934

Last year of model B and C singles. New Art Deco styling touches added to the bikes, including stylized handlebars and motif and streamlined taillights. Production rises to around 10,000 bikes, almost tripling previous year's output.

MODEL B

• ENGINE FLATHEAD SINGLE

• CAPACITY 21cu. in.

MODEL C

• ENGINE FLATHEAD SINGLE

• CAPACITY 30.5cu. in.

MODEL VARIATIONS

C/CB

MODEL R

• ENGINE FLATHEAD V-twin

• CAPACITY 45cu. in.

MODEL VARIATIONS

R/RL/RLD/RLDX/RLX/RS/RSX/RX

MODEL G SERVI-CAR

• ENGINE FLATHEAD V-twin

• CAPACITY 45cu. in.

MODEL VARIATIONS

G/GA/GD/GDT/GE

MODEL V

• ENGINE FLATHEAD V-twin

• CAPACITY 74cu. in.

MODEL VARIATIONS

V/VD/VLD/VDS/VFD/VFDS

• NEW FEATURES
AIRFLOW TAIL LIGHT
DIAMOND-SHAPED LOGO ON TANK

• COLORS
SINGLES IN SILVER AND RED OR
OLIVE GREEN AND BLACK. TWINS IN
A RANGE OF SIX STANDARD COLORS.

1935

New 80cu. in. unit available for model V, using same chassis as on 74cu. in. model. Joe Petrali wins every round of dirt-track championship.

MODEL R

• ENGINE FLATHEAD V-twin

• CAPACITY 45cu. in.

MODEL VARIATIONS

R/RL/RLD/RLDR/RS/RSR

1935 RL

MODEL G SERVI-CAR

• ENGINE Flathead V-twin

• CAPACITY 45cu. in.

MODEL VARIATIONS

G/GA/GD/GDT/GE

MODEL V

• ENGINE Flathead V-twins

• CAPACITY 74cu. in. & 80cu. in.

MODEL VARIATIONS

VD/VLD/VLDJ/VDS/VFD/VFDS
(74cu. in.)
VLDD/VDDS (80cu. in.)

• NEW FEATURES
CONSTANT-MESH THREE-SPEED
TRANSMISSION ON MODEL R
NEW GEARBOX FOR MODELS R AND G
BEEHIVE TAIL LIGHT

• COLORS
A RANGE OF SIX TWO-TONE COLORS
AS STANDARD. TWO OPTIONAL
COLORS AVAILABLE AT EXTRA COST.

1936

New 61cu. in. OHV "Knucklehead"
V-twin engine on new model E. Last
year of model V flathead big-twins.

MODEL R

• ENGINE Flathead V-twin

• CAPACITY 45cu. in.

MODEL VARIATIONS

R/RL/RLD/RLDR/RS/RSR

MODEL G SERVI-CAR

• ENGINE Flathead V-twin

• CAPACITY 45cu. in.

MODEL VARIATIONS

G/GA/GD/GDT/GE

MODEL E

• ENGINE OHV V-twin

• CAPACITY 61cu. in.

MODEL VARIATIONS

E/EL/ES

1936 61EL

MODEL V

• ENGINE Flathead V-twins

• CAPACITY 74cu. in. & 80cu. in.

MODEL VARIATIONS

VD/VLD/VMG/VDS/VFD/VFDS
(74cu. in.)
VLH/VHS/VFH/VFHS (80cu. in.)

• NEW FEATURES
DRY SUMP LUBRICATION (MODEL E)
FOUR-SPEED GEARBOX

(MODEL E & OPTION ON MODEL V)
BULLET AND CIRCLE LOGO
(ALL MODELS)

• COLORS
A RANGE OF FIVE TWO-TONE COLORS
AS STANDARD.

1937

Revised flathead engines for new
models W and U. All models restyled
to follow the look of the model E.
Dry sump recirculating oil system
now on all models.

MODEL W

• ENGINE Flathead V-twin

• CAPACITY 45cu. in.

MODEL VARIATIONS

W/WL/WLD/WLDR/WS/WSR

MODEL G SERVI-CAR

• ENGINE Flathead V-twin

• CAPACITY 45cu. in.

MODEL VARIATIONS

G/GA/GD/GDT/GE

MODEL E

• ENGINE OHV V-twin

• CAPACITY 61cu. in.

• MODEL VARIATIONS

E/EL/ES

MODEL U

• ENGINE Flathead V-twin

• CAPACITY 74cu. in. & 80cu. in.

MODEL VARIATIONS

U/UL/US/UMG (74cu. in.)
UH/ULH/UHS (80cu. in.)

• NEW FEATURES
DRY SUMP LUBRICATION (NOW ON
ALL MODELS)
INSTRUMENT PANEL ON FUEL TANK
(AS PER 1936 MODEL E)

• COLORS
A RANGE OF COLORS BECAME
AVAILABLE AS STANDARD.

1938

Revisions to all models. All bikes
ordered with an option package.
Joe Petrali retires from racing.

MODEL W

• ENGINE Flathead V-twin

• CAPACITY 45cu. in.

MODEL VARIATIONS

W/WL/WLD/WLDR/WS

MODEL G SERVI-CAR

• ENGINE Flathead V-twin

• CAPACITY 45cu. in.

MODEL VARIATIONS

G/GA/GD/GDT/GE

MODEL E

• ENGINE OHV V-twin

• CAPACITY 61cu. in.

MODEL VARIATIONS

EL/ES

MODEL U

• ENGINE Flathead V-twins

• CAPACITY 74cu. in. & 80cu. in.

MODEL VARIATIONS

U/UL/US/UMG (74cu. in.)
UH/ULH/UHS (80cu. in.)

• COLORS
A RANGE OF SIX COLORS AS
STANDARD.

1939

Another year of modifications across
the ranges. Body of the Servi-Car is
enlarged. First WLA Army model
shipped to Fort Knox.

MODEL W

• ENGINE Flathead V-twin

• CAPACITY 45cu. in.

MODEL VARIATIONS

W/WL/WLD/WLDR/WS

MODEL G SERVI-CAR

• ENGINE Flathead V-twin

• CAPACITY 45cu. in.

MODEL VARIATIONS

G/GA/GD/GDT/GE

MODEL E

• ENGINE OHV V-twin

• CAPACITY 61cu. in.

• MODEL VARIATIONS

EL/ES

MODEL U

• ENGINE Flathead V-twins

• CAPACITY 74cu. in. & 80cu. in.

• MODEL VARIATIONS

U/UL/US/UMG (74cu. in.)
UH/ULH/UHS (80cu. in.)

• NEW FEATURES
"BOAT-TAIL" TAILLIGHT
"CAT'S-EYE" INSTRUMENT CONSOLE

• COLORS
A RANGE OF FOUR COLORS AS
STANDARD.

1940

First full production year of WLA,
military version of model W. Tank
badges chrome-plated across the range.

MODEL W

• ENGINE Flathead V-twin

• CAPACITY 45cu. in.

MODEL VARIATIONS

W/WL/WLD/WLDR/WS/WLA

MODEL G SERVI-CAR

• ENGINE Flathead V-twin

• CAPACITY 45cu. in.

MODEL VARIATIONS

G/GA/GD/GDT

MODEL E

• ENGINE OHV V-twin

• CAPACITY 61cu. in.

MODEL VARIATIONS

EL/ES

MODEL U

• ENGINE Flathead V-twins

• CAPACITY 74cu. in. & 80cu. in.

MODEL VARIATIONS

U/UL/US/UMG (74cu. in.)
UH/ULH/UHS (80cu. in.)

• NEW FEATURES
CHROME-PLATED TANK NAMEPLATE

• COLORS A RANGE OF FIVE COLORS
AS STANDARD.

1941

First year of new OHV 74 cu. in.
model F. New model UA is army
version of model U flathead. Civilian
bike production suspended during the
year. Last year of 80cu. in. big-twins.

MODEL W

• ENGINE Flathead V-twin

• CAPACITY 45cu. in.

MODEL VARIATIONS

WL/WLS/WLD/WLDR/
WR/WLA/WLC

1941 WLA

MODEL G SERVI-CAR

• ENGINE Flathead V-twin

• CAPACITY 45cu. in.

MODEL VARIATIONS

G/GA/GD/GDT

MODEL E

• ENGINE OHV V-twin

• CAPACITY 61cu. in.

MODEL VARIATIONS

E/EL/ES

MODEL F

• ENGINE OHV V-twin

• CAPACITY 74cu. in.

MODEL VARIATIONS

F/FL/FS

MODEL U

- ENGINE Flathead V-twins
- CAPACITY 74cu. in. & 80cu. in.

MODEL VARIATIONS

U/UL/US/UA (74cu. in.)
UH/ULH/UHS (80cu. in.)

- NEW FEATURES
Revised "aircraft-style"
speedometer

- COLORS
A range of five colors as
standard.

1942

First year of new shaft-drive model XA
twin for military use only. Blackout
lights and other military modifications
feature on WLA. Death of Walter
Davidson, one the founding brothers
and President of the company.

MODEL W

- ENGINE Flathead V-twin
- CAPACITY 45cu. in.

MODEL VARIATIONS

WL/WLS/WLD/WLA/WLC

1942 WLA

MODEL G SERVI-CAR

- ENGINE Flathead V-twin
- CAPACITY 45cu. in.

MODEL VARIATIONS

G/GA

MODEL XA

- ENGINE Flathead
SHAFT-DRIVE OPPOSED TWIN
- CAPACITY 45cu. in.

1942 XA

MODEL E

- ENGINE OHV V-twin
- CAPACITY 61cu. in.

MODEL VARIATIONS

E/EL/ES/ELC

MODEL F

- ENGINE OHV V-twin
- CAPACITY 74cu. in.

MODEL VARIATIONS

F/FL/FS

MODEL U

- ENGINE Flathead V-twin
- CAPACITY 74cu. in.

MODEL VARIATIONS

U/UL/US

- COLORS A range of five colors
as standard.

1943

Model WLA accounts for virtually all
Harley production, with most sent
abroad. Last year of XA.

MODEL W

- ENGINE Flathead V-twin
- CAPACITY 45cu. in.

MODEL VARIATIONS

WLA/WLC

MODEL G SERVI-CAR

- ENGINE Flathead V-twin
- CAPACITY 45cu. in.

MODEL VARIATIONS

G/GA

MODEL XA

- ENGINE Flathead shaft-drive
OPPOSED TWIN
- CAPACITY 45cu. in.

MODEL E

- ENGINE OHV V-twin
- CAPACITY 61cu. in.

MODEL VARIATIONS

E/EL/ES

MODEL F

- ENGINE OHV V-twin
- CAPACITY 74cu. in.

MODEL VARIATIONS

F/FL/FS

MODEL U

- ENGINE Flathead V-twin
- CAPACITY 74cu. in.

MODEL VARIATIONS

U/UL/US

- NEW FEATURES
"Winged-face" horn

- COLORS
Gray or Silver.

1944

Bulk of production again for the
military, with only a very few bikes
manufactured for civilians or
the police.

MODEL W

- ENGINE Flathead V-twin
- CAPACITY 45cu. in.

MODEL VARIATIONS

WL/WLA/WLC/WSR

MODEL G SERVI-CAR

- ENGINE Flathead V-twin
- CAPACITY 45cu. in.

MODEL VARIATIONS

G/GA

MODEL E

- ENGINE OHV V-twin
- CAPACITY 61cu. in.

MODEL VARIATIONS

E/EL/ES

MODEL F

- ENGINE OHV V-twin
- CAPACITY 74 cu. in.

MODEL VARIATIONS

F/FL/FS

MODEL U

- ENGINE Flathead V-twin
- CAPACITY 74cu. in.

MODEL VARIATIONS

U/UL/US

- COLORS
Gray or Silver.

1944 U

1945

Civilian production resumes in
November, with options packages now
available once more. No chrome
finishes on any models.

MODEL W

- ENGINE Flathead V-twin
- CAPACITY 45cu. in.

MODEL VARIATIONS

WL/WLA/WSR

MODEL G SERVI-CAR

- ENGINE Flathead V-twin
- CAPACITY 45cu. in.

MODEL VARIATIONS

G/GA

MODEL E

- ENGINE OHV V-twin
- CAPACITY 61cu. in.

MODEL VARIATIONS

E/EL/ES

MODEL F

- ENGINE OHV V-twin
- CAPACITY 74cu. in.

MODEL VARIATIONS

F/FL/FS

MODEL U

- ENGINE Flathead V-twin
- CAPACITY 74cu. in.

MODEL VARIATIONS

U/UL/US

- COLORS Gray.

1946

US Government sells off thousands of
surplus WLAs, but WL model remains
the most popular bike.

MODEL W

- ENGINE Flathead V-twin
- CAPACITY 45cu. in.

MODEL VARIATIONS

WL/WL-SP

MODEL G SERVI-CAR

- ENGINE Flathead V-twin
- CAPACITY 45cu. in.

MODEL VARIATIONS

G/GA

MODEL E

- ENGINE OHV V-twin
- CAPACITY 61cu. in.

MODEL VARIATIONS

E/EL/ES

MODEL F

- ENGINE OHV V-twin
- CAPACITY 74cu. in.

MODEL VARIATIONS

F/FL/FS

MODEL U

- ENGINE Flathead V-twin
- CAPACITY 74cu. in.

MODEL VARIATIONS

U/UL/US

- COLORS
Gray or Flight Red.

1947

Harley acquires rights to produce German DKW 125cc machine. First year of Harley's model S lightweight. Last year of "Knucklehead" V-twin. New manufacturing plant opens on Capitol Drive, in Milwaukee.

MODEL S

- ENGINE AIR-COOLED TWO-STROKE SINGLE
- CAPACITY 125cc

MODEL W

- ENGINE FLATHEAD V-TWIN
- CAPACITY 45cu. IN.

MODEL VARIATIONS
WL/WL-SP

MODEL G SERVI-CAR

- ENGINE FLATHEAD V-TWIN
- CAPACITY 45cu. IN.

MODEL VARIATIONS
G/GA

MODEL E

- ENGINE OHV V-TWIN
- CAPACITY 61cu. IN.

MODEL VARIATIONS
E/EL/ES

MODEL F

- ENGINE OHV V-TWIN
- CAPACITY 74cu. IN.

MODEL VARIATIONS
F/FL/FS

MODEL U

- ENGINE FLATHEAD V-TWIN
- CAPACITY 74cu. IN.

MODEL VARIATIONS
U/UL/US

- NEW FEATURES
"Tombstone" tail light
New red ball Harley emblem

- COLORS
RANGE OF FIVE COLORS NOW AVAILABLE AGAIN AS STANDARD.

1948

New OHV "Panhead" engine with hydraulic valve lifters for models E and F big-twins. Last year of model U flathead V-twin.

MODEL S

- ENGINE AIR-COOLED TWO-STROKE SINGLE
- CAPACITY 125cc

MODEL W

- ENGINE FLATHEAD V-TWIN
- CAPACITY 45cu. IN.

MODEL VARIATIONS
WL/WLS/WR/WL-SP

MODEL G SERVI-CAR

- ENGINE FLATHEAD V-TWIN
- CAPACITY 45cu. IN.

MODEL VARIATIONS
G/GA

MODEL E

- ENGINE OHV V-TWIN
- CAPACITY 61cu. IN.

MODEL VARIATIONS
E/EL/ES

MODEL F

- ENGINE OHV V-TWIN
- CAPACITY 74cu. IN.

MODEL VARIATIONS
F/FL/FS

MODEL U

- ENGINE FLATHEAD V-TWIN
- CAPACITY 74cu. IN.

MODEL VARIATIONS
U/UL/US

- NEW FEATURES
"Panhead" engine

- COLORS
RANGE OF FOUR COLORS AVAILABLE AS STANDARD.

1949

Hydraulic front forks on models E and F big-twins give birth to the Hydra-Glide name. Leading-link forks still available as an option.

MODEL S

- ENGINE AIR-COOLED TWO-STROKE SINGLE
- CAPACITY 125cc

MODEL W

- ENGINE FLATHEAD V-TWIN
- CAPACITY 45cu. IN.

MODEL VARIATIONS
WL/WLS/WR/WL-SP

1949 WR DIRT TRACK

MODEL G SERVI-CAR

- ENGINE FLATHEAD V-TWIN
- CAPACITY 45cu. IN.

MODEL VARIATIONS
G/GA

MODEL E

- ENGINE OHV V-TWIN
- CAPACITY 61cu. IN.

MODEL VARIATIONS
E/ES/EL/ELP/EP

MODEL F

- ENGINE OHV V-TWIN
- CAPACITY 74cu. IN.

MODEL VARIATIONS
F/FL/FS/FP/FLP

- NEW FEATURES
HYDRAULIC FORKS ON MODELS E & F

- COLORS
BLACK ONLY FOR SINGLE. RANGE OF FOUR STANDARD COLORS FOR TWINS, WITH ONE OPTIONAL COLOR.

1950

Models FL, S, and EL are manufactured in largest quantities. Death of Arthur Davidson, last of the original four founders. Safety bars optional on Hydra-Glides.

MODEL S

- ENGINE AIR-COOLED TWO-STROKE SINGLE
- CAPACITY 125cc

MODEL W

- ENGINE FLATHEAD V-TWIN
- CAPACITY 45cu. IN.

MODEL VARIATIONS
WL/WLA/WLS/WR/WL-SP

MODEL G SERVI-CAR

- ENGINE FLATHEAD V-TWIN
- CAPACITY 45cu. IN.

MODEL VARIATIONS
G/GA

MODEL E

- ENGINE OHV V-TWIN
- CAPACITY 61cu. IN.

MODEL VARIATIONS
E/ES/EL

MODEL F

- ENGINE OHV V-TWIN
- CAPACITY 74cu. IN.

MODEL VARIATIONS
F/FL/FS/

- NEW FEATURES
REDESIGNED CYLINDER HEADS ON "PANHEAD" ENGINE

- COLORS
RANGE OF FOUR STANDARD COLORS, WITH FOUR EXTRAS AVAILABLE FOR TWINS.

1951

Last year of model WL. Model S gets telescopic forks and is known as the Tele-Glide S.

MODEL S

- ENGINE AIR-COOLED TWO-STROKE SINGLE
- CAPACITY 125cc

MODEL W

- ENGINE FLATHEAD V-TWIN
- CAPACITY 45cu. IN.

MODEL VARIATIONS
WL/WLA/WLS/WR/WRTT/WL-SP

MODEL G SERVI-CAR

- ENGINE FLATHEAD V-TWIN
- CAPACITY 45cu. IN.

MODEL VARIATIONS
G/GA

MODEL E

- ENGINE OHV V-TWIN
- CAPACITY 61cu. IN.

MODEL VARIATIONS
EL/ELS

MODEL F

- ENGINE OHV V-TWIN
- CAPACITY 74cu. IN.

MODEL VARIATIONS
FL/FLS

1951 74FL

- NEW FEATURES
REDESIGNED HARLEY SCRIPT ON TANK

- COLORS
RANGE OF THREE COLORS AND TWO EXTRAS FOR SINGLES. THREE STANDARD AND THREE EXTRAS FOR TWINS.

1952

First year of new 45cu. in. flathead model K Sport. Last year of 61cu. in. model E, model S 125, and the few remaining model Ws. Last year that special sidecar models are listed.

MODEL S

- ENGINE AIR-COOLED TWO-STROKE SINGLE
- CAPACITY 125cc

MODEL W

- ENGINE Flathead V-twin
- CAPACITY 45 cu. in.

MODEL VARIATIONS

WR/WLS

MODEL G SERVI-CAR

- ENGINE Flathead V-twin
- CAPACITY 45cu. in.

MODEL VARIATIONS

G/GA

MODEL K

- ENGINE Flathead V-twin
- CAPACITY 45cu. in.

MODEL VARIATIONS

K

1952 K

RACING MODELS

KR/KRTT

MODEL E

- ENGINE OHV V-twin
- CAPACITY 61cu. in.

MODEL VARIATIONS

EL/ELS/ELF

MODEL F

- ENGINE OHV V-twin
- CAPACITY 74cu. in.

MODEL VARIATIONS

FL/FLS/FLF

- NEW FEATURES
45cu. in. unit construction engine on model K
Optional foot shift on models E and F (ELF and FLF)

- COLORS
Range of three colors with one optional for singles, five colors with three optional for twins, and three colors with one optional for model K.

1953

First year of ST, Harley's single now uprated to 165cc. Last year of model K. The Indian Motorcycle Company ceases trading, leaving Harley-Davidson as the only major US manufacturer.

MODEL ST

- ENGINE Air-cooled two-stroke single
- CAPACITY 165cc

MODEL G SERVI-CAR

- ENGINE Flathead V-twin
- CAPACITY 45cu. in.

MODEL VARIATIONS

G/GA

MODEL K

- ENGINE Flathead V-twin
- CAPACITY 45cu. in.

MODEL VARIATIONS

K

RACING MODELS

KK/KR/KRTT (racers)

MODEL F

- ENGINE OHV V-twin
- CAPACITY 74cu. in.

MODEL VARIATIONS

FL/FLE/FLEF/FLF

- NEW FEATURES
New Stewart-Warner "edge-lighted" speedometer on twins

- COLORS
Range of three colors with one extra for single, five colors with three extra for twins.

1954

In an attempt to improve performance, new 55cu. in. flathead V-twin model KH replaces model K. Model S returns with two-strokes now called Hummers. Old 45cu. in. unit retained in racers.

MODEL S

- ENGINE Air-cooled two-stroke single
- CAPACITY 125cc

MODEL ST

- ENGINE Air-cooled two-stroke single
- CAPACITY 165cc

MODEL VARIATIONS

ST/STU

MODEL G SERVI-CAR

- ENGINE Flathead V-twin
- CAPACITY 45cu. in.

MODEL VARIATIONS

G/GA

MODEL KH

- ENGINE Flathead V-twin
- CAPACITY 55cu. in.

MODEL VARIATIONS

KH

RACING MODELS

KHRM (racer)
KR/KRTT (45cu. in. racers)

MODEL F

- ENGINE OHV V-twin
- CAPACITY 74cu. in.

MODEL VARIATIONS

FL/FLE/FLEF/FLF

- NEW FEATURES
55cu. in. engine on model KH
Dual exhaust system an option on model F Hydra-Glides

- COLORS
Range of 15 solid and two-tone colors as standard.

1955

First year of Super Sport FLH and FLHF models added to the Hydra-Glide range.

MODEL S

- ENGINE Air-cooled two-stroke single
- CAPACITY 125cc

MODEL ST

- ENGINE Air-cooled two-stroke single
- CAPACITY 165cc

MODEL VARIATIONS

ST/STU

1955 ST

MODEL G SERVI-CAR

- ENGINE Flathead V-twin
- CAPACITY 45cu. in.

MODEL VARIATIONS

G/GA

MODEL KH

- ENGINE Flathead V-twin
- CAPACITY 55cu. in.

MODEL VARIATIONS

KH/KHK

RACING MODELS

KHRM/KHRM/KHRTT
KR/KRTT (45cu. in.)

MODEL F

- ENGINE OHV V-twin
- CAPACITY 74cu. in.

MODEL VARIATIONS

FL/FLE/FLEF/FLF/FLH/FLHF

- NEW FEATURES
Tank emblem with letter "V" and Harley-Davidson in script behind

- COLORS
Range of seven solid colors as standard. Two-tone combinations at extra cost.

1956

Last year of KH production bikes. Elvis Presley appears on the cover of *The Enthusiast* magazine sitting on a Harley-Davidson.

MODEL S

- ENGINE Air-cooled two-stroke single
- CAPACITY 125cc

MODEL ST

- ENGINE Air-cooled two-stroke single
- CAPACITY 165cc

MODEL VARIATIONS

ST/STU

MODEL G SERVI-CAR

- ENGINE Flathead V-twin
- CAPACITY 45cu. in.

MODEL VARIATIONS

G/GA

MODEL KH

- ENGINE Flathead V-twin
- CAPACITY 55cu. in.

MODEL VARIATIONS

KH/KHK

RACING MODELS

KHRTT (racer)
KR/KRTT (45cu. in. racers)

MODEL F

- ENGINE OHV V-twin
- CAPACITY 74cu. in.

MODEL VARIATIONS

FL/FLE/FLEF/FLF/FLH/FLHF

- NEW FEATURES
Round taillight on Hummer
"Victory" camshaft on FLH

- COLORS
Range of six solid colors as standard. Metallic Green available at extra cost on twins.

1957

First year of XL Sportster, based on model KH. It will go on to become the longest-running model in motorcycle history. The model K series now consists of racing bikes only.

MODEL S

- ENGINE Air-cooled two-stroke single
- CAPACITY 125cc

MODEL ST

- ENGINE Air-cooled two-stroke single
- CAPACITY 165cc

MODEL VARIATIONS

ST/STU

MODEL G SERVI-CAR

- ENGINE Flathead V-twin
- CAPACITY 45cu. in.

MODEL VARIATIONS

G/GA

MODEL XL

- ENGINE OHV V-twin
- CAPACITY 55cu. in.

1957 XL SPORTSTER

MODEL F

- ENGINE OHV V-twin
- CAPACITY 74cu. in.

MODEL VARIATIONS

FL/FLF/FLH/FLHF

MODEL K RACERS

- ENGINE Flathead V-twins
- CAPACITY 45cu. in. & 55cu. in.

MODEL VARIATIONS

Models KR/KRTT (45cu. in. Flathead V-twins)
Model KHRTT (55cu. in. Flathead V-twin)

- NEW FEATURES
OHV V-twin unit construction engine on XL
New tank logo (designed by Willie G. Davidson) has red Harley-Davidson script over silver and red quadrants

- COLORS
Range of seven colors as standard.

1958

Expanded XL range. Rear suspension on model F big-twins sees Harley's first Duo-Glides.

MODEL S

- ENGINE Air-cooled two-stroke single
- CAPACITY 125cc

MODEL ST

- ENGINE Air-cooled two-stroke single
- CAPACITY 165cc

MODEL VARIATIONS

ST/STU

MODEL G SERVI-CAR

- ENGINE Flathead V-twin
- CAPACITY 45cu. in.

MODEL VARIATIONS

G/GA

MODEL XL

- ENGINE OHV V-twin
- CAPACITY 55cu. in.

MODEL VARIATIONS

XL/XLH/XLC/XLCH

RACING MODELS

XLRTT

MODEL F

- ENGINE OHV V-twin
- CAPACITY 74cu. in.

MODEL VARIATIONS

FL/FLF/FLH/FLHF

- NEW FEATURES
Rear shock absorbers on model F, now dubbed the Duo-Glide

MODEL KR RACERS

- ENGINE Flathead V-twin
- CAPACITY 45cu. in.

MODEL VARIATIONS

KR/KRTT

- COLORS
WIDE RANGE OF COLORS, ALL AS STANDARD.

1959

Last year of models S and ST. Also last year of XL in standard Sportster form.

MODEL S

- ENGINE Air-cooled two-stroke single
- CAPACITY 125cc

MODEL ST

- ENGINE Air-cooled two-stroke single
- CAPACITY 165cc

MODEL VARIATIONS

ST/STU

MODEL G SERVI-CAR

- ENGINE Flathead V-twin
- CAPACITY 45cu. in.

MODEL VARIATIONS

G/GA

MODEL XL

- ENGINE OHV V-twin
- CAPACITY 55cu. in.

MODEL VARIATIONS

XL/XLH/XLC/XLCH

RACING MODELS

XLRTT

MODEL F

- ENGINE OHV V-twin
- CAPACITY 74cu. in.

MODEL VARIATIONS

FL/FLF/FLH/FLHF

MODEL KR RACERS

- ENGINE Flathead V-twin
- CAPACITY 45cu. in.

MODEL VARIATIONS

KR/KRTT

- NEW FEATURES
"Arrow-flight" tank badge

- COLORS
Range of seven colors as standard.

1960

Harley-Davidson tries to entice scooter buyers with its new Topper scooter. Model BT replaces ST with upgraded 165cc engine. Harley buys a controlling stake in Italian small-bike manufacturer, Aermacchi.

MODEL A TOPPER

- ENGINE Air-cooled two-stroke single
- CAPACITY 165cc

MODEL VARIATIONS

A/AU

MODEL BT SUPER 10

- ENGINE Air-cooled two-stroke single
- CAPACITY 165cc

MODEL VARIATIONS

BT/BTU

MODEL G SERVI-CAR

- ENGINE Flathead V-twin
- CAPACITY 45cu. in.

MODEL VARIATIONS

G/GA

MODEL XL

- ENGINE OHV V-twin
- CAPACITY 55cu. in.

MODEL VARIATIONS

XLH/XLCH

RACING MODELS

XLRTT

MODEL F

- ENGINE OHV V-twin
- CAPACITY 74cu. in.

MODEL VARIATIONS

FL/FLF/FLH/FLHF

1960 FLH Duo-Glide

MODEL KR RACERS

- ENGINE Flathead V-twin
- CAPACITY 45cu. in.

MODEL VARIATIONS

KR/KRTT

- NEW FEATURES
165cc engine on models A and BT

- COLORS
Range of colors as standard. Special paint finishes at extra cost.

1961

First year for Aermacchi-derived Sprint, last year for BT Super 10. Dual exhaust optional on Duo-Glides.

MODEL AH TOPPER

- ENGINE Air-cooled two-stroke single
- CAPACITY 165cc

MODEL VARIATIONS

AH/AH

MODEL BT SUPER 10

- ENGINE Air-cooled two-stroke single
- CAPACITY 165cc

MODEL VARIATIONS

BT/BTU

SPRINT

- ENGINE Horizontal OHV four-stroke single
- CAPACITY 250cc

MODEL VARIATIONS

C

RACING MODELS

CRTT

MODEL G SERVI-CAR

- ENGINE Flathead V-twin
- CAPACITY 45cu. in.

MODEL VARIATIONS

G/GA

MODEL XL

- ENGINE OHV V-twin
- CAPACITY 55cu. in.

MODEL VARIATIONS

XLH/XLCH

RACING MODELS

XLRTT

MODEL F

- ENGINE OHV V-TWIN
- CAPACITY 74CU. IN.

MODEL VARIATIONS

FL/FLF/FLH/FLHF

MODEL KR RACERS

- ENGINE Flathead V-TWIN
- CAPACITY 45CU. IN.

MODEL VARIATIONS

KR/KRTT

- NEW FEATURES
New circle and star tank emblem

- COLORS Range of colors as standard. Special paint finishes at extra cost.

1962

Single range expanded to include 175cc models. Last year of 165cc in BT range. Harley-Davidson buys 60 percent stake in Tomahawk Boat Company for fiberglass components.

MODEL AH TOPPER

- ENGINE Air-cooled two-stroke single
- CAPACITY 165cc

MODEL VARIATIONS

AH/AU

MODEL B SINGLES

- ENGINE Air-cooled two-stroke singles
- CAPACITY 165cc & 175cc

MODEL VARIATIONS

BTU Pacer/BTF Ranger (165cc)
BT Pacer/BTH Scat(175cc)

SPRINT

- ENGINE Horizontal OHV four-stroke single
- CAPACITY 250cc

MODEL VARIATIONS

C/H

RACING MODELS

CRTT

MODEL G SERVI-CAR

- ENGINE Flathead V-TWIN
- CAPACITY 45CU. IN.

MODEL VARIATIONS

G/GA

MODEL XL

- ENGINE OHV V-TWIN
- CAPACITY 55CU. IN.

MODEL VARIATIONS

XLH/XLCH

RACING MODELS

XLRTT

MODEL F

- ENGINE OHV V-TWIN
- CAPACITY 74CU. IN.

MODEL VARIATIONS

FL/FLF/FLH/FLHF

MODEL KR RACERS

- ENGINE Flathead V-TWIN
- CAPACITY 45CU. IN.

MODEL VARIATIONS

KR/KRTT

- NEW FEATURES
New 175cc single engine "Tombstone" speedometer featured on twins

- COLORS
Range of colors as standard. Special paint finishes available at extra cost.

1963

F models are biggest sellers this year. Willie G. Davidson joins the Harley-Davidson board.

MODEL AH TOPPER

- ENGINE Air-cooled two-stroke single
- CAPACITY 165cc

MODEL VARIATIONS

AH/AU

MODEL B SINGLES

- ENGINE Air-cooled two-stroke single
- CAPACITY 175cc

MODEL VARIATIONS

BT Pacer/BTU Pacer/BTH Scat

SPRINT

- ENGINE Horizontal OHV four-stroke single
- CAPACITY 250cc

MODEL VARIATIONS

C/H

RACING MODELS

CRTT

MODEL G SERVI-CAR

- ENGINE Flathead V-TWIN
- CAPACITY 45CU. IN.

MODEL VARIATIONS

G/GA

MODEL XL

- ENGINE OHV V-TWIN
- CAPACITY 55CU. IN.

MODEL VARIATIONS

XLH/XLCH

RACING MODELS

XLRTT

MODEL F

- ENGINE OHV V-TWIN
- CAPACITY 74CU. IN.

MODEL VARIATIONS

FL/FLF/FLH/FLHF

MODEL KR RACERS

- ENGINE Flathead V-TWIN
- CAPACITY 45CU. IN.

MODEL VARIATIONS

KR/KRTT

- NEW FEATURES
Revised tank emblem

- COLORS
Range of colors as standard. Special paint finishes available at extra cost.

1964

Last year of AU, restricted version of Topper. Last year of Duo-Glide. Servi-Car gets Harley's first electric starter.

MODEL AH TOPPER

- ENGINE Air-cooled two-stroke single
- CAPACITY 165cc

MODEL VARIATIONS

AH/AU

1964 AH TOPPER

MODEL B SINGLES

- ENGINE Air-cooled two-stroke single
- CAPACITY 175cc

MODEL VARIATIONS

BT Pacer/BTU Pacer/BTH Scat

SPRINT

- ENGINE Horizontal OHV four-stroke single
- CAPACITY 250cc

MODEL VARIATIONS

C/H

RACING MODELS

CRTT

MODEL GE SERVI-CAR

- ENGINE Flathead V-TWIN
- CAPACITY 45CU. IN.

MODEL XL

- ENGINE OHV V-TWIN
- CAPACITY 55CU. IN.

MODEL VARIATIONS

XLH/XLCH

RACING MODELS

XLRTT

MODEL F

- ENGINE OHV V-TWIN
- CAPACITY 74CU. IN.

MODEL VARIATIONS

FL/FLF/FLH/FLHF

MODEL KR RACERS

- ENGINE Flathead V-TWIN
- CAPACITY 45CU. IN.

MODEL VARIATIONS

KR/KRTT

- NEW FEATURES
Electric starter on model GE Servi-Car

- COLORS
Range of four colors as standard. Special paint finishes at extra cost.

1965

Electric starter on model F gives birth to the Electra Glide name. New model M 50cc Italian bikes. Last year of "Panhead" engine, Topper scooter, and Scat and Pacer singles.

MODEL M LEGGERO

- ENGINE Air-cooled two-stroke single
- CAPACITY 50cc

MODEL AH TOPPER

- ENGINE Air-cooled single
- CAPACITY 165cc

MODEL B SINGLE

- ENGINE Air-cooled two-stroke single
- CAPACITY 175cc

MODEL VARIATIONS

BT Pacer/BTH Scat

SPRINT

- ENGINE Horizontal OHV four-stroke single
- CAPACITY 250cc

MODEL VARIATIONS

C/H

RACING MODELS

CRTT

MODEL GE SERVI-CAR

- ENGINE Flathead V-TWIN
- CAPACITY 45CU. IN.

MODEL XL

- **ENGINE** OHV V-twin
- **CAPACITY** 55cu. in.

MODEL VARIATIONS

XLH/XLCH

RACING MODELS

XLRTT

MODEL F

- **ENGINE** OHV V-twin
- **CAPACITY** 74cu. in.

MODEL VARIATIONS

FLHB/FLHFB/FLB/FLFB

MODEL KR RACERS

- **ENGINE** Flathead V-twin
- **CAPACITY** 45cu. in.

MODEL VARIATIONS

KR/KRTT

- **NEW FEATURES**
ELECTRIC STARTER ON MODEL F

- **COLORS**
RANGE OF COLORS AS STANDARD.
SPECIAL PAINT FINISHES AVAILABLE
AT EXTRA COST.

1966

First year of new "Shovelhead" OHV engine, a revision of the "Panhead" unit. Only year of Bobcat. Last year of 50cc in M series and US-made singles.

MODEL M LEGGERO

- **ENGINE** Air-cooled two-stroke single
- **CAPACITY** 50cc

MODEL VARIATIONS

M-50/M-50 Sport

1966 M-50 Sport

MODEL BTH BOBCAT

- **ENGINE** Air-cooled two-stroke single
- **CAPACITY** 175cc

SPRINT

- **ENGINE** Horizontal OHV four-stroke single
- **CAPACITY** 250cc

MODEL VARIATIONS

C/H

RACING MODELS

CRTT

MODEL GE SERVI-CAR

- **ENGINE** Flathead V-twin
- **CAPACITY** 45cu. in.

MODEL XL

- **ENGINE** OHV V-twin
- **CAPACITY** 55cu. in.

MODEL VARIATIONS

XLH/XLCH

RACING MODELS

XLRTT

MODEL F

- **ENGINE** OHV V-twin
- **CAPACITY** 74cu. in.

MODEL VARIATIONS

FLHB/FLHFB/FLB/FLFB

MODEL KR RACERS

- **ENGINE** Flathead V-twin
- **CAPACITY** 45cu. in.

MODEL VARIATIONS

KR/KRTT

- **NEW FEATURES**
"Shovelhead" engine for model F
HEXAGONAL TANK BADGE

- **COLORS**
RANGE OF FOUR COLORS AS
STANDARD. SPECIAL PAINT FINISHES
AT EXTRA COST.

1967

New version of model M added with larger-capacity engine. XLH model gets an electric starter.

MODEL M LEGGERO

- **ENGINE** Air-cooled two-stroke single
- **CAPACITY** 65cc

MODEL VARIATIONS

M-65/M-65 Sport

SPRINT

- **ENGINE** Horizontal OHV four-stroke single
- **CAPACITY** 250cc

MODEL VARIATIONS

SS/H

RACING MODELS

CRTT

1967 CRTT

MODEL GE SERVI-CAR

- **ENGINE** Flathead V-twin
- **CAPACITY** 45cu. in.

MODEL XL

- **ENGINE** OHV V-twin
- **CAPACITY** 55cu. in.

MODEL VARIATIONS

XLH/XLCH

MODEL F

- **ENGINE** OHV V-twin
- **CAPACITY** 74cu. in.

MODEL VARIATIONS

FLHB/FLHFB/FLB/FLFB

- **COLORS**
BLACK OR BLUE ON TWINS.
OTHER COLORS AVAILABLE
AT EXTRA COST.

1968

M model has 125cc version added. Last year for 250cc Sprint. Cal Raybourn rides a KRTT to victory at Daytona 200. First year of no kick-starter on Sportsters.

MODEL M

- **ENGINE** Air-cooled two-stroke singles
- **CAPACITY** 65cc & 125cc

MODEL VARIATIONS

M-65/M-65 Sport,
M-125 Rapido

SPRINT

- **ENGINE** Horizontal OHV four-stroke single
- **CAPACITY** 250cc

MODEL VARIATIONS

SS/H

MODEL GE SERVI-CAR

- **ENGINE** Flathead V-twin
- **CAPACITY** 45cu. in.

MODEL XL

- **ENGINE** OHV V-twin
- **CAPACITY** 55cu. in.

MODEL VARIATIONS

XLH/XLCH

MODEL F

- **ENGINE** OHV V-twin
- **CAPACITY** 74cu. in.

MODEL VARIATIONS

FLHB/FLHFB/FLB/FLFB

- **NEW FEATURES**
125cc engine in M-Rapido
ELECTRIC STARTER AND RESTYLED
INSTRUMENT CONSOLE ON
SPORTSTERS

- **COLORS**
BLACK OR ORANGE ON TWINS.
OTHER COLORS AVAILABLE AT
EXTRA COST.

1969

American Machine and Foundry Company (AMF) buys controlling stake in Harley-Davidson. First year of 350cc Sprint.

MODEL M

- **ENGINE** Air-cooled two-stroke singles
- **CAPACITY** 65cc & 125cc

MODEL VARIATIONS

M-65/M-65 Sport, M-125 Rapido

SPRINT

- **ENGINE** Horizontal OHV four-stroke single
- **CAPACITY** 350cc

MODEL VARIATIONS

SS/ERS

MODEL GE SERVI-CAR

- **ENGINE** Flathead V-twin
- **CAPACITY** 45cu. in.

1969 GE SERVI-CAR

MODEL XL

- **ENGINE** OHV V-twin
- **CAPACITY** 55cu. in.

MODEL VARIATIONS

XLH/XLCH

MODEL F

- **ENGINE** OHV V-twin
- **CAPACITY** 74cu. in.

MODEL VARIATIONS

FLHB/FLHFB/FLB/FLFB

- **NEW FEATURES**
350cc ENGINE IN SPRINT

- **COLORS**
BLACK, ORANGE, OR WHITE ON
TWINS. OTHER COLORS AVAILABLE
AT EXTRA COST.

1970

First year of new 45cu. in. racer, the XR750, which goes on to be Harley's most successful racing bike. Also first year of "Shovelhead" with alternator rather than generator and the 100cc Aermacchi Baja.

MODEL M

- **ENGINE** Air-cooled two-stroke singles
- **CAPACITY** 65cc, 100cc & 125cc

MODEL VARIATIONS

M-65 LEGERRO (65cc), MSR BAJA
(100cc), MLS RAPIDO (125cc)

SPRINT

- ENGINE HORIZONTAL OHV
FOUR-STROKE SINGLE

- CAPACITY 350cc

MODEL VARIATIONS

SS/ERS

MODEL GE SERV-CAR

- ENGINE FLATHEAD V-TWIN

- CAPACITY 45CU. IN.

MODEL XR

- ENGINE OHV V-TWIN

- CAPACITY 45CU. IN.

MODEL XL

- ENGINE OHV V-TWIN

- CAPACITY 55CU. IN.

MODEL VARIATIONS

XLH/XLCH

MODEL F

- ENGINE OHV V-TWIN

- CAPACITY 74CU. IN.

MODEL VARIATIONS

FLH/FLHF/FLP/FLPF

- NEW FEATURES "SHOVELHEAD"
UNIT NOW HAS ALTERNATOR INSTEAD
OF GENERATOR
NEW RACING V-TWIN FOR XR750

- COLORS
WHITE OR BLUE.

1971

First year of new FX Super Glide,
brainchild of Willie G. Davidson and
Harley's first official custom bike. Last
year of 55cu. in. Sportster and M-65.
XLCH is the most popular
model this year.

MODEL M

- ENGINE AIR-COOLED TWO-
STROKE SINGLES

- CAPACITY 65cc, 100cc, & 125cc

MODEL VARIATIONS

M-65 LEGERRO (65cc), MSR BAJA
(100cc), MLS RAPIDO (125cc)

SPRINT

- ENGINE HORIZONTAL OHV
FOUR-STROKE SINGLE

- CAPACITY 350cc

MODEL VARIATIONS

SS/SX/ERS

MODEL GE SERV-CAR

- ENGINE FLATHEAD V-TWIN

- CAPACITY 45CU. IN.

MODEL XR

- ENGINE OHV V-TWIN

- CAPACITY 45CU. IN.

MODEL XL

- ENGINE OHV V-TWIN

- CAPACITY 55CU. IN.

MODEL VARIATIONS

XLH/XLCH

MODEL F

- ENGINE OHV V-TWIN

- CAPACITY 74CU. IN.

MODEL VARIATIONS

FLH/FLHF/FLP/FLPF

MODEL FX

- ENGINE OHV V-TWIN

- CAPACITY 74CU. IN.

1971 FX

- NEW FEATURES
AMF LOGO NOW ADDED TO TANK

- COLORS
WHITE OR BLUE.

1972

First year for 61cu. in. Sportster and
last year for Baja and Rapido singles.
AMF adds its name to tank logos
across the range. Harley enters into
production of snowmobiles.

MODEL M

- ENGINE AIR-COOLED TWO-
STROKE SINGLES

- CAPACITY 65cc, 100cc,
& 125cc

MODEL VARIATIONS

M-65 LEGERRO/MC-65 SPORTSTER
(65cc), MSR BAJA (100cc)
MLS RAPIDO (125cc)

SPRINT

- ENGINE HORIZONTAL OHV
FOUR-STROKE SINGLE

- CAPACITY 350cc

MODEL VARIATIONS

SS/SX/ERS

MODEL GE SERV-CAR

- ENGINE FLATHEAD V-TWIN

- CAPACITY 45CU. IN.

MODEL XR RACER

- ENGINE OHV V-TWIN

- CAPACITY 45CU. IN.

MODEL XRTT RACER

- ENGINE OHV V-TWIN

- CAPACITY 45CU. IN.

1972 XRTT

MODEL XL

- ENGINE OHV V-TWIN

- CAPACITY 61CU. IN.

MODEL VARIATIONS

XLH/XLCH

MODEL F

- ENGINE OHV V-TWIN

- CAPACITY 74CU. IN.

MODEL VARIATIONS

FLH/FLHF/FLP/FLPF

MODEL FX

- ENGINE OHV V-TWIN

- CAPACITY 74CU. IN.

- NEW FEATURES
RECTANGULAR AMF/HD BADGE

- COLORS
WHITE OR BLUES.

1973

Last year of model GE Servi-Car, with
disc brakes added. New range of 90cc
and 125cc singles introduced.

TWO-STROKE SINGLES

- ENGINE AIR-COOLED TWO-
STROKE SINGLES

- CAPACITY 90cc & 125cc

MODEL VARIATIONS

X-90/Z-90 (90cc)
125SS/125SX (125cc)

SPRINT

- ENGINE HORIZONTAL OHV
FOUR-STROKE SINGLE

- CAPACITY 350cc

MODEL VARIATIONS

SS/SX

MODEL GE SERV-CAR

- ENGINE FLATHEAD V-TWIN

- CAPACITY 45CU. IN.

MODEL XR

- ENGINE OHV V-TWIN

- CAPACITY 45CU. IN.

MODEL XL

- ENGINE OHV V-TWIN

- CAPACITY 61CU. IN.

MODEL VARIATIONS

XLH/XLCH

MODEL F

- ENGINE OHV V-TWIN

- CAPACITY 74CU. IN.

MODEL VARIATIONS

FLH/FLHF/FLP/FLPF

MODEL FX

- ENGINE OHV V-TWIN

- CAPACITY 74CU. IN.

- NEW FEATURES
90cc SINGLE ENGINE
125cc SINGLE ENGINE

- COLORS
WHITE OR BLUE.

1974

First year for 175cc singles and
electric start Super Glide. Harley wins
250cc World Championship with
Aermacchi machine. Last year for
Sprint and 90cc bikes.

TWO-STROKE SINGLES

- ENGINE AIR-COOLED TWO-
STROKE SINGLES

- CAPACITY 90cc, 125cc,
AND 175cc

MODEL VARIATIONS

X-90/Z-90 (90cc)
125SS/125SX (125cc)
175SS/175SX (175cc)

SPRINT

- ENGINE HORIZONTAL OHV
FOUR-STROKE SINGLE

- CAPACITY 350cc

MODEL VARIATIONS

SS/SX

MODEL XR

- ENGINE OHV V-TWIN

- CAPACITY 45CU. IN.

MODEL XL

- ENGINE OHV V-TWIN

- CAPACITY 61CU. IN.

MODEL VARIATIONS

XLH/XLCH

MODEL F

- ENGINE OHV V-TWIN

- CAPACITY 74CU. IN.

MODEL VARIATIONS

FL/FLH/FLHF

MODEL FX

- ENGINE OHV V-TWIN

- CAPACITY 74CU. IN.

MODEL VARIATIONS

FX/FXE

- **NEW FEATURES**
ELECTRIC STARTER NOW AVAILABLE ON SUPER GLIDE (FXE)

- **COLORS**
WHITE, BURGUNDY, BLACK, OR BLUE.

1975

First year for new 250cc two-stroke singles. Chassis manufacture and final bike assembly moved to larger premises in York, Pennsylvania, with engine manufacture kept in Milwaukee.

TWO-STROKE SINGLES

- **ENGINE** AIR-COOLED TWO-STROKE SINGLES

- **CAPACITY** 125cc, 175cc, & 250cc

MODEL VARIATIONS

125SS/125SX (125cc)
175SS/175SX (175cc)
250SS/250SX (250cc)

MODEL XR

- **ENGINE** OHV V-TWIN

- **CAPACITY** 45CU. IN.

MODEL XL

- **ENGINE** OHV V-TWIN

- **CAPACITY** 61CU. IN.

MODEL VARIATIONS

XLH/XLCH

MODEL F

- **ENGINE** OHV V-TWIN

- **CAPACITY** 74CU. IN.

MODEL VARIATIONS

FL/FLH/FLHF

MODEL FX

- **ENGINE** OHV V-TWIN

- **CAPACITY** 74CU. IN.

MODEL VARIATIONS

FX/FXE

- **COLORS**
WHITE, BURGUNDY, BLACK, OR BLUE.

1976

Only year that the World Championship-winning RR250 is cataloged. Production rises to almost 50,000 bikes this year.

TWO-STROKE SINGLES

- **ENGINE** AIR-COOLED TWO-STROKE SINGLES

- **CAPACITY** 125cc, 175cc, & 250cc

MODEL VARIATIONS

125SS/125SX (125cc)
175SS/175SX (175cc)
250SS/250SX (250cc)

MODEL RR250

- **ENGINE** TWO-STROKE TWIN

- **CAPACITY** 250cc

1976 RR250

MODEL XRTT

- **ENGINE** OHV V-TWIN

- **CAPACITY** 45CU. IN.

MODEL XL

- **ENGINE** OHV V-TWIN

- **CAPACITY** 61CU. IN.

MODEL VARIATIONS

XLH/XLCH (PLUS SPECIAL "LIBERTY" EDITIONS)

MODEL F

- **ENGINE** OHV V-TWIN

- **CAPACITY** 74CU. IN.

MODEL VARIATIONS

FLH/FLH "LIBERTY" EDITION

MODEL FX

- **ENGINE** OHV V-TWIN

- **CAPACITY** 74CU. IN.

MODEL VARIATIONS

FX/FXE (PLUS "LIBERTY" EDITIONS)

- **COLORS**
WHITE, BURGUNDY, BLACK, ORANGE, SILVER, BROWN, OR BLUE.

1977

First year of XLCR Café Racer, FXS Low Rider, and FLT. Only year of FLHS Electra Glide. Last year of Harley's singles.

TWO-STROKE SINGLES

- **ENGINE** AIR-COOLED TWO-STROKE SINGLES

- **CAPACITY** 125cc, 175cc, & 250cc

MODEL VARIATIONS

125SS/125SX (125cc)
175SS/175SX (175cc)
250SS/250SX (250cc)

MODEL MX RACER

- **ENGINE** TWO-STROKE SINGLE

- **CAPACITY** 250cc

MODEL XR

- **ENGINE** OHV V-TWIN

- **CAPACITY** 45CU. IN.

MODEL XL

- **ENGINE** OHV V-TWIN

- **CAPACITY** 61CU. IN.

MODEL VARIATIONS

XLH/XLCH/XLT/XLCR

MODEL F

- **ENGINE** OHV V-TWIN

- **CAPACITY** 74CU. IN.

MODEL VARIATIONS

FLH/FLHS

MODEL FX

- **ENGINE** OHV V-TWIN

- **CAPACITY** 74CU. IN.

MODEL VARIATIONS

FX/FXE/FXS

- **NEW FEATURES**
REDESIGNED FRAME FOR NEW XLCR

- **COLORS**
CHOICE OF EIGHT STANDARD COLORS.

1978

First year of new 80cu. in. "Shovelhead" unit. Last year of XLCR, XLT, and FX Super Glide. Harley ends its association with the Aermacchi company.

MODEL MX RACER

- **ENGINE** TWO-STROKE SINGLE

- **CAPACITY** 250cc

MODEL XR

- **ENGINE** OHV V-TWIN

- **CAPACITY** 45CU. IN.

MODEL XL

- **ENGINE** OHV V-TWIN

- **CAPACITY** 61CU. IN.

MODEL VARIATIONS

XLH/XLH/XLCH/XLT/XLCR
XLH (ANNIVERSARY EDITION)

1978 XLCR

MODEL F

- **ENGINE** OHV V-TWINS

- **CAPACITY** 74CU. IN. & 80CU. IN.

MODEL VARIATIONS

FLH/FLH (ANNIVERSARY EDITION)
FLH (80CU. IN.)

MODEL FX

- **ENGINE** OHV V-TWIN

- **CAPACITY** 74CU. IN.

MODEL VARIATIONS

FX/FXE/FXS

- **NEW FEATURES**
80CU. IN. BIG-TWIN ENGINE
DUAL DISC FRONT BRAKES ON SPORTSTERS

- **COLORS**
CHOICE OF NINE STANDARD COLORS.

1979

First year of new XLS "roadster" model, FXEF Fat Bobs, FLHC Electra Glide Classic, and FXS Low Rider with larger 80cu. in. engine. Last year of XLCH.

MODEL XR

- **ENGINE** OHV V-TWIN

- **CAPACITY** 45CU. IN.

MODEL XL

- **ENGINE** OHV V-TWIN

- **CAPACITY** 61CU. IN.

MODEL VARIATIONS

XLH/XLCH/XLS

MODEL F

- **ENGINE** OHV V-TWINS

- **CAPACITY** 74CU. IN. & 80CU. IN.

MODEL VARIATIONS

FLH/FLH(POLICE)(74CU. IN.)
FLH/FLH(POLICE)/FLHC (80CU. IN.)

MODEL FX

- **ENGINE** OHV V-TWINS

- **CAPACITY** 74CU. IN. & 80CU. IN.

MODEL VARIATIONS

FX/FXE/FXS/FXEF (74CU. IN.)
FX/FXS/FXEF (80CU. IN.)

- **COLORS**
CHOICE OF NINE STANDARD COLORS.

1980

Last year of 74cu. in. "Shovelhead" unit. New models FLT Tour Glide, FXB Sturgis, FXWG Wide Glide, and XLH Hugger.

MODEL XR

- **ENGINE** OHV V-TWIN

- **CAPACITY** 45CU. IN.

MODEL XL

- **ENGINE** OHV V-TWIN

- **CAPACITY** 61CU. IN.

MODEL VARIATIONS

XLH/XLS

MODEL FLH

- **ENGINE** OHV V-TWINS

- **CAPACITY** 74CU. IN. & 80CU. IN.

MODEL VARIATIONS

FLH (74CU. IN.)
FLH/FLHS/FLHC (80CU. IN.)

MODEL FLT
- ENGINE OHV V-TWIN
- CAPACITY 80CU. IN.

MODEL FX
- ENGINE OHV V-TWINS
- CAPACITY 74CU. IN. & 80CU. IN.

MODEL VARIATIONS
FXE/FXS (74CU. IN.)
FXB/FXS/FXEF/FXWG
(80CU. IN.)

- NEW FEATURES
RUBBER-MOUNTED ENGINE TO
REDUCE VIBRATION ON FLT
FIVE-SPEED GEARBOX ON FLT
BELT FINAL-DRIVE ON FXB

- COLORS
CHOICE OF EIGHT STANDARD
COLORS.

1981

Harley-Davidson management buys
company back from AMF. Advertising
slogan is "The Eagle Soars Alone".
First year for 80cu. in. FXE Super
Glide, replacing 74cu. in. model.

MODEL XR
- ENGINE OHV V-TWIN
- CAPACITY 45CU. IN.

MODEL XL
- ENGINE OHV V-TWIN
- CAPACITY 61CU. IN.

MODEL VARIATIONS
XLH/XLS

MODEL FLH
- ENGINE OHV V-TWIN
- CAPACITY 80CU. IN.

MODEL VARIATIONS
FLH/FLH (HERITAGE)/
FLHS/FLHC

MODEL FLT
- ENGINE OHV V-TWIN
- CAPACITY 80CU. IN.

MODEL VARIATIONS
FLT/FLTC

MODEL FX
- ENGINE OHV V-TWIN
- CAPACITY 80CU. IN.

MODEL VARIATIONS
FXE/FXS/FXB/FXEF/FXWG

- NEW FEATURES
FLH HERITAGE IS RETURNED
TO ORIGINAL STYLING WITH
FEATURES SUCH AS LEATHER
SADDLEBAGS

- COLORS
CHOICE OF 12 STANDARD COLORS.

1982

Despite streamlining of the company
resulting in a number of layoffs, new
FXR and FXRS Super Glides
launched in exciting new era for the
company. Last year for Fat Bob, Low
Rider, and Sturgis. The Sportster
celebrates its 25th anniversary, with
special editions released.

MODEL XR
- ENGINE OHV V-TWIN
- CAPACITY 45CU. IN.

MODEL XL
- ENGINE OHV V-TWIN
- CAPACITY 61CU. IN.

MODEL VARIATIONS
XLH/XLS, XLHA/XLSA
(ANNIVERSARY EDITIONS)

MODEL FLH
- ENGINE OHV V-TWIN
- CAPACITY 80CU. IN.

MODEL VARIATIONS
FLH/FLHS/FLHF

MODEL FLT
- ENGINE OHV V-TWIN
- CAPACITY 80CU. IN.

MODEL VARIATIONS
FLT/FLTC

MODEL FX
- ENGINE OHV V-TWIN
- CAPACITY 80CU. IN.

MODEL VARIATIONS
FXE/FXS/FXB/FXWG

MODEL FXR
- ENGINE OHV V-TWIN
- CAPACITY 80CU. IN.

MODEL VARIATIONS
FXR/FXRS

- NEW FEATURES NEW FRAME ON
XL MODELS WITH REPOSITIONED OIL
TANK AND BATTERY

- COLORS
CHOICE OF 13 STANDARD COLORS.

1983

New model XR1000 sports bike based
on Sportster engine and XLH chassis.
Various new styles and models
released based on 80cu. in. engine,
including FXSB Low Rider and FXDG
with rear disc wheel. Harley Owners'
Group (H.O.G.) formed.

MODEL XL/XR
- ENGINE OHV V-TWIN
- CAPACITY 61CU. IN.

MODEL VARIATIONS
XLH/XLS/XLX-61/XR1000

MODEL FLH
- ENGINE OHV V-TWIN
- CAPACITY 80CU. IN.

MODEL VARIATIONS
FLH/FLHS

MODEL FLT
- ENGINE OHV V-TWIN
- CAPACITY 80CU. IN.

MODEL VARIATIONS
FLT/FLTC/FLHT/FLHTC

MODEL FX
- ENGINE OHV V-TWIN
- CAPACITY 80CU. IN.

MODEL VARIATIONS
FXE/FXSB/FXDG/FXWG

MODEL FXR
- ENGINE OHV V-TWIN
- CAPACITY 80CU. IN.

MODEL VARIATIONS
FXR/FXRS/FXRT

- NEW FEATURES
ALUMINUM DISC REAR WHEEL ON
FXDG
BELT FINAL-DRIVE ON MOST MODELS

- COLORS
CHOICE OF NINE STANDARD COLORS.

1984

Last year of "Shovelhead" and first
year of all-new V2 "Evolution"
engine. New FXST Softail is an
update of "hardtail" bikes, with rear
suspension incorporated behind frame.
Only year of limited-edition FXRDG,
based on FXRS, with rear disc wheel.
Last year of XR1000. Sportsters switch
to single front disc brake.

MODEL XL/XR
- ENGINE OHV V-TWIN
- CAPACITY 61CU. IN.

MODEL VARIATIONS
XLH/XLS/XLX-61/XR1000

1984 XR1000

MODEL FLH
- ENGINE OHV V-TWIN
(SHOVELHEAD)
- CAPACITY 80CU. IN.

MODEL VARIATIONS
FLH/FLHS/FLHX

1984 FLHX

MODEL FLT
- ENGINE OHV V-TWIN
(EVOLUTION)
- CAPACITY 80CU. IN.

MODEL VARIATIONS
FLTC/FLHTC

MODEL FX
- ENGINE OHV V-TWIN
(SHOVELHEAD)
- CAPACITY 80CU. IN.

MODEL VARIATIONS
FXE/FXSB/FXWG

MODEL FXR
- ENGINE OHV V-TWIN
(EVOLUTION)
- CAPACITY 80CU. IN.

MODEL VARIATIONS
FXRS/FXRT/FXRP/FXRDG

MODEL FXST
- ENGINE OHV V-TWIN
(EVOLUTION)
- CAPACITY 80CU. IN.

- NEW FEATURES
FXST SOFTAIL FEATURES REAR
SUSPENSION HIDDEN UNDER FRAME

- COLORS
CHOICE OF NINE STANDARD COLORS.
[SOME EARLIER BIKES MARKED AS
HAVING THE EVOLUTION ENGINE
MAY WELL HAVE USED THE
SHOVELHEAD UNIT.]

1985

First and only year for FXRC Low
Glide Custom as well as FXEF and
FXSB Fat Bobs. Evolution engine now
used across the big-twin range. Belt
drive on most models.

MODEL XL
- ENGINE OHV V-TWIN
- CAPACITY 61CU. IN.

MODEL VARIATIONS
XLH/XLS/XLX-61

MODEL FLT
- ENGINE OHV V-TWIN
- CAPACITY 80CU. IN.

MODEL VARIATIONS

FLTC/FLHTC

MODEL FXWG

- **ENGINE** OHV V-twin
- **CAPACITY** 80cu. in.

MODEL FXR

- **ENGINE** OHV V-twin
- **CAPACITY** 80cu. in.

MODEL VARIATIONS
FXRS/FXRT/FXRC

MODEL FXS

- **ENGINE** OHV V-twin
- **CAPACITY** 80cu. in.

MODEL VARIATIONS
FXST/FXSB/FXEF

- **COLORS**
CHOICE OF SIX STANDARD COLORS.

1986

Revamp of the Sportster range, with new 883cc and 1100cc Evolution engines. Special "Liberty" editions of some models introduced, with special graphics. New FLST Heritage Softail in style of old Hydra-Glide, with wire-spoked wheels and shrouds on front forks. FXR reintroduces the Super Glide name, with chain drive for this year only.

MODEL XLH

- **ENGINE** OHV V-twins
- **CAPACITY** 883cc & 1100cc

MODEL VARIATIONS
XLH-883 (3 styles)/XLH-1100
XLH-1100 (LIBERTY EDITION)

MODEL FLH

- **ENGINE** OHV V-twin
- **CAPACITY** 80cu. in.

MODEL VARIATIONS
FLHTC

MODEL FLT

- **ENGINE** OHV V-twin
- **CAPACITY** 80cu. in.

MODEL VARIATIONS
FLTC

MODEL FXR

- **ENGINE** OHV V-twin
- **CAPACITY** 80cu. in.

MODEL VARIATIONS
FXR/FXRS/FXRT/FXRD

MODEL FXS

- **ENGINE** OHV V-twin
- **CAPACITY** 80cu. in.

MODEL VARIATIONS
FXST/FXSTC/FXWG/FLST

- **NEW FEATURES**
883cc AND 1100cc EVOLUTION ENGINES FOR SPORTSTER RANGE
- **COLORS**
CHOICE OF SEVEN STANDARD COLORS. SPECIAL ANNIVERSARY EDITIONS OF SOME MODELS.

1987

Harley-Davidson floats on the stock market. First year of FLHS Electra Glide Sport and FLSTC, with special touring accessories. Super Glides get belt final drive. Harley celebrates 30th anniversary of Sportster and renews contracts with police departments.

MODEL XLH

- **ENGINE** OHV V-twins
- **CAPACITY** 883cc & 1100cc

MODEL VARIATIONS
XLH-883 (3 styles)/XLH-1100
XLH-1100 (ANNIVERSARY edition)

MODEL FLH

1987 XLH-883

- **ENGINE** OHV V-twin
- **CAPACITY** 80cu. in.

MODEL VARIATIONS
FLHS/FLHTC

MODEL FLT

- **ENGINE** OHV V-twin
- **CAPACITY** 80cu. in.

MODEL VARIATIONS
FLTC

MODEL FXR

- **ENGINE** OHV V-twin
- **CAPACITY** 80cu. in.

MODEL VARIATIONS
FXR/FXRS/FXRT/FXRD

SOFTAILS

- **ENGINE** OHV V-twin
- **CAPACITY** 80cu. in.

MODEL VARIATIONS
FXST/FXSTC/FLST/FLSTC

- **NEW FEATURES**
DETACHABLE WINDSHIELD ON FLHS
- **COLORS**
CHOICE OF SEVEN STANDARD COLORS. ANNIVERSARY EDITIONS OF SOME MODELS.

1988

Larger 1200cc unit replaces 1100cc engine in Sportster range. First year of Springer Softail, deploying leading-link forks.

MODEL XLH

- **ENGINE** OHV V-twins
- **CAPACITY** 883cc & 1200cc

MODEL VARIATIONS
XLH-883 (3 styles)/XLH-1200

MODEL FLH

- **ENGINE** OHV V-twin
- **CAPACITY** 80cu. in.

MODEL VARIATIONS
FLHS/FLHTC

MODEL FLT

- **ENGINE** OHV V-twin
- **CAPACITY** 80cu. in.

MODEL VARIATIONS
FLTC

MODEL FXR

- **ENGINE** OHV V-twin
- **CAPACITY** 80cu. in.

MODEL VARIATIONS
FXR/FXRS/FXRT/FXLR

SOFTAILS

- **ENGINE** OHV V-twin
- **CAPACITY** 80cu. in.

MODEL VARIATIONS
FXST/FXSTC/FXSTS/FLST/
FLSTC

- **NEW FEATURES**
1200cc SPORTSTER ENGINE SPRINGER leading-link forks RETURN ON THE FXST
- **COLORS**
CHOICE OF 12 STANDARD COLORS. SOME 85TH ANNIVERSARY MODELS ALSO PRODUCED.

1989

New racing series begins based solely on the 883 Sportster. Ultra package available for Electra Glide.

MODEL XLH

- **ENGINE** OHV V-twins
- **CAPACITY** 883cc & 1200cc

MODEL VARIATIONS
XLH-883 (3 styles)/XLH-1200

MODEL FLH

- **ENGINE** OHV V-twin
- **CAPACITY** 80cu. in.

MODEL VARIATIONS
FLHS/FLHTC/FLHTCU

MODEL FLT

- **ENGINE** OHV V-twin
- **CAPACITY** 80cu. in.

MODEL VARIATIONS
FLTC/FLTCU

![1989 FLTC motorcycle]
1989 FLTC

MODEL FXR

- **ENGINE** OHV V-twin
- **CAPACITY** 80cu. in.

MODEL VARIATIONS
FXR/FXRS/FXRT/FXLR

SOFTAILS

- **ENGINE** OHV V-twin
- **CAPACITY** 80cu. in.

MODEL VARIATIONS
FXST/FXSTC/FXSTS/FLST/
FLSTC

- **NEW FEATURES**
"ULTRA" ACCESSORY PACKAGE (WITH U DESIGNATION) OFFERS CRUISE CONTROL AND MUSIC SYSTEM
- **COLORS**
CHOICE OF 11 STANDARD COLORS.

1990

First year of FLSTF "Fat Boy", with disc wheels, dual exhaust system, and unique Harley styling. Self-canceling turn signals now on all models.

MODEL XLH

- **ENGINE** OHV V-twins
- **CAPACITY** 883cc & 1200cc

MODEL VARIATIONS
XLH-883 (3 styles)/XLH-1200

MODEL FLH

- **ENGINE** OHV V-twin
- **CAPACITY** 80 cu. in.

MODEL VARIATIONS
FLHS/FLHTC/FLHTCU

MODEL FLT

- **ENGINE** OHV V-twin
- **CAPACITY** 80 cu. in.

MODEL VARIATIONS
FLTC/FLTCU/

MODEL FXR

- **ENGINE** OHV V-twin
- **CAPACITY** 80 cu. in.

MODEL VARIATIONS

FXR/FXRS/FXRT/FXLR

SOFTAILS

- **ENGINE** OHV V-twin
- **CAPACITY** 80 cu. in.

MODEL VARIATIONS

FXST/FXSTC/FXSTS/FLST/
FLSTC/FLSTF

- **NEW FEATURES**
Disc wheels on FLSTF
- **COLORS**
Choice of 12 standard colors.

1991

First and only year of FXDB Dyna-Glide Sturgis, using new Dyna-Glide frame with rubber engine mountings.

MODEL XLH

- **ENGINE** OHV V-twins
- **CAPACITY** 883cc & 1200cc

MODEL VARIATIONS

XLH-883 (3 styles)/XLH-1200

MODEL FLH

- **ENGINE** OHV V-twin
- **CAPACITY** 80 cu. in.

MODEL VARIATIONS

FLHS/FLHTC/FLHTCU

MODEL FLT

- **ENGINE** OHV V-twin
- **CAPACITY** 80 cu. in.

MODEL VARIATIONS

FLTC/FLTCU

MODEL FXR

- **ENGINE** OHV V-twin
- **CAPACITY** 80 cu. in.

MODEL VARIATIONS

FXR/FXRS/FXRT/FXLR

MODEL FXD

- **ENGINE** OHV V-twin
- **CAPACITY** 80cu. in.

SOFTAILS

- **ENGINE** OHV V-twin
- **CAPACITY** 80cu. in.

MODEL VARIATIONS

FXSTC/FXSTS/FLSTC/FLSTF

- **NEW FEATURES**
DYNA CHASSIS WITH LARGE SECTION SPINE AND VIBRATION-ISOLATING RUBBER MOUNTING SYSTEM FIVE-SPEED GEARBOX ON SPORTSTERS, SOME WITH BELT DRIVE
- **COLORS**
CHOICE OF 11 STANDARD COLORS.

1992

First (and only) year of FXDB Dyna-Glide Daytona and FXDC Super Dyna-Glide Custom.

MODEL XLH

- **ENGINE** OHV V-twin
- **CAPACITY** 883cc & 1200cc

MODEL VARIATIONS

XLH-883 (3 styles)/XLH-1200

MODEL FLH

- **ENGINE** OHV V-twin
- **CAPACITY** 80cu. in.

MODEL VARIATIONS

FLHS/FLHTC/FLHTCU

MODEL FLT

- **ENGINE** OHV V-twin
- **CAPACITY** 80cu. in.

MODEL VARIATIONS

FLTC/FLTCU

MODEL FXR

- **ENGINE** OHV V-twin
- **CAPACITY** 80cu. in.

MODEL VARIATIONS

FXR/FXRS/FXRT/FXLR

MODEL FXD

- **ENGINE** OHV V-twin
- **CAPACITY** 80cu. in.

MODEL VARIATIONS

FXDB/FXDC

SOFTAILS

- **ENGINE** OHV V-twin
- **CAPACITY** 80cu. in.

MODEL VARIATIONS

FXSTC/FXSTS/FLSTC/FLSTF

- **NEW FEATURES** Gold wheels and rear sprocket on FXDB
- **COLORS**
WIDE CHOICE OF STANDARD COLORS AVAILABLE.

1993

Harley buys a stake in Buell sports bike company. Return of the Wide Glide in the form of the new FXDWG. New FXDL Low Rider, also based on Dyna-Glide chassis. FLSTN is limited-edition Nostalgia version of Heritage Softail.

MODEL XLH

- **ENGINE** OHV V-twin
- **CAPACITY** 883cc & 1200cc

MODEL VARIATIONS

XLH-883 (3 styles)/XLH-1200/
XLH-1200(anniversary model)

MODEL FLH

- **ENGINE** OHV V-twin
- **CAPACITY** 80cu. in.

MODEL VARIATIONS

FLHS/FLHTC/FLHTCU

MODEL FLT

- **ENGINE** OHV V-twin
- **CAPACITY** 80cu. in.

MODEL VARIATIONS

FLTC/FLTCU

MODEL FXR

- **ENGINE** OHV V-twin
- **CAPACITY** 80cu. in.

MODEL VARIATIONS

FXR/FXRS/FXLR

MODEL FXD

- **ENGINE** OHV V-twin
- **CAPACITY** 80cu. in.

MODEL VARIATIONS

FXDL/FXDWG

SOFTAILS

- **ENGINE** OHV V-twin
- **CAPACITY** 80cu. in.

MODEL VARIATIONS

FXSTC/FXSTS/FLSTC/FLSTF/
FLSTN

- **COLORS**
CHOICE OF 10 STANDARD COLORS. SPECIAL ANNIVERSARY EDITIONS OF SOME MODELS.

1994

Last year of FXR models. First year of FLHR Road King and FXDS Convertible, with detachable saddlebags allowing it to convert from a tourer to a custom bike.

MODEL XLH

- **ENGINE** OHV V-twins
- **CAPACITY** 883cc & 1200cc

MODEL VARIATIONS

XLH-883 (3 styles)/XLH-1200

MODEL FLH

- **ENGINE** OHV V-twin
- **CAPACITY** 80cu. in.

MODEL VARIATIONS

FLHR/FLHTC/FLHTCU

MODEL FLT

- **ENGINE** OHV V-twin
- **CAPACITY** 80cu. in.

MODEL VARIATIONS

FLTC/FLTCU

MODEL FXR

- **ENGINE** OHV V-twin
- **CAPACITY** 80cu. in.

MODEL VARIATIONS

FXR/FXLR

MODEL FXD

- **ENGINE** OHV V-twin
- **CAPACITY** 80cu. in.

MODEL VARIATIONS

FXDL/FXDWG/FXDS

SOFTAILS

- **ENGINE** OHV V-twin
- **CAPACITY** 80cu. in.

MODEL VARIATIONS

FXSTC/FXSTS/FLSTC/FLSTF/
FLSTN

- **COLORS**
CHOICE OF FIVE STANDARD COLORS.

1995

Harley-Davidson's first fuel-injected model, the FLHTCUI. Electra Glide celebrates its 30th anniversary, with some special editions produced. New base models for both the Electra Glide and the Super Glide, plus a new Bad Boy with Springer forks.

MODEL XLH

- **ENGINE** OHV V-twins
- **CAPACITY** 883cc & 1200cc

MODEL VARIATIONS

XLH-883 (3 styles)/XLH-1200

MODEL FLH

- **ENGINE** OHV V-twin
- **CAPACITY** 80cu. in.

MODEL VARIATIONS

FLHR/FLHT/FLHTC/FLHTCU/
FLHTCUI (Anniversary edition)

MODEL FLT

- **ENGINE** OHV V-twin
- **CAPACITY** 80cu. in.

MODEL VARIATIONS

FLTC/FLTCU

MODEL FXD

- **ENGINE** OHV V-twin
- **CAPACITY** 80cu. in.

MODEL VARIATIONS

FXD/FXDL/FXDWG/FXDS

SOFTAILS

- **ENGINE** OHV V-twin
- **CAPACITY** 80cu. in.

MODEL VARIATIONS

FXSTC/FXSTS/FXSTSB/FLSTC/
FLSTF/FLSTN

- **NEW FEATURES**
SEQUENTIAL PORT FUEL-INJECTION ON FLHTCUI
ADJUSTABLE SUSPENSION ON FLHT MODELS

- **COLORS**
CHOICE OF 11 STANDARD COLORS.

1996

Revised Sportster range sees expanded 1200cc models including Custom and Sport bikes. Custom has laced front wheel and disc rear wheel. Fuel injection now on a number of Electra Glide models.

MODEL XLH

- **ENGINE** OHV V-TWINS
- **CAPACITY** 883cc & 1200cc

MODEL VARIATIONS

XLH-883 (2 STYLES)
XLH-1200/XLH-1200C/XLH-1200S

MODEL FLH

- **ENGINE** OHV V-TWIN
- **CAPACITY** 80CU. IN.

MODEL VARIATIONS

FLHR/FLHRI

MODEL FLHT

- **ENGINE** OHV V-TWIN
- **CAPACITY** 80CU. IN.

MODEL VARIATIONS

FLHT/FLHTC/FLHTCU/FLHTCUI

MODEL FXD

- **ENGINE** OHV V-TWIN
- **CAPACITY** 80CU. IN.

MODEL VARIATIONS

FXD/FXDL/FXDWG/FXDS

SOFTAILS

- **ENGINE** OHV V-TWIN
- **CAPACITY** 80CU. IN.

MODEL VARIATIONS

FXSTC/FXSTS/FXSTSB/FLSTC/
FLSTF/FLSTN

- **COLORS**
CHOICE OF TEN STANDARD COLORS.

1997

Very little change in model range, but new FLSTS Heritage Springer Softail revisits late-1940s styling.

MODEL XLH

- **ENGINE** OHV V-TWINS
- **CAPACITY** 883cc & 1200cc

MODEL VARIATIONS

XLH-883 (2 STYLES)
XLH-1200/XLH-1200C/XLH-1200S

MODEL FLH

- **ENGINE** OHV V-twin
- **CAPACITY** 80cu. in.

MODEL VARIATIONS

FLHR/FLHRI

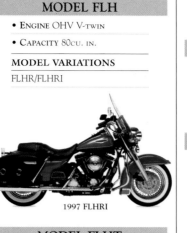

1997 FLHRI

MODEL FLHT

- **ENGINE** OHV V-TWIN
- **CAPACITY** 80CU. IN.

MODEL VARIATIONS

FLHT/FLHTC/FLHTCU/FLHTCUI

MODEL FXD

- **ENGINE** OHV V-TWIN
- **CAPACITY** 80CU. IN.

MODEL VARIATIONS

FXD/FXDL/FXDWG/FXDS

SOFTAILS

- **ENGINE** OHV V-TWIN
- **CAPACITY** 80CU. IN.

MODEL VARIATIONS

FXSTC/FXSTS/FLSTC/
FLSTF/FLSTS

- **COLORS**
WIDE CHOICE OF STANDARD COLORS.

1998

New Sportster Sport released with chrome and silver finish and digital single-fire ignition system. Rumors of the release of Harley's new big-twin engine after five years in development.

MODEL XLH

- **ENGINE** OHV V-TWINS
- **CAPACITY** 883cc & 1200cc

MODEL VARIATIONS

XLH-883 (2 STYLES)
XLH-1200/XLH-1200C/
XLH-1200S

MODEL FLH

- **ENGINE** OHV V-TWIN
- **CAPACITY** 80 CU. IN.

MODEL VARIATIONS

FLHR/FLHRCI

MODEL FLHT

- **ENGINE** OHV V-TWIN
- **CAPACITY** 80CU. IN.

MODEL VARIATIONS

FLHT/FLHTCI/FLHTCUI

MODEL FXD

- **ENGINE** OHV V-TWIN
- **CAPACITY** 80CU. IN.

MODEL VARIATIONS

FXD/FXDL/FXDWG/FXDS

SOFTAILS

- **ENGINE** OHV V-TWIN
- **CAPACITY** 80CU. IN.

MODEL VARIATIONS

FXSTC/FXSTS/FLSTS/FLSTC/
FLSTF

- **COLORS**
WIDE RANGE OF STANDARD COLORS.

1999

Introduction of Harley's largest ever engine, the Twin Cam 88 on Dyna and FLH/T models. Increased engine displacement with increased bore and shorter stroke create an estimated 10 percent power increase over the Evolution.

MODEL XL/XLH

- **ENGINE** OHV V-TWINS
- **CAPACITY** 883cc & 1200cc

MODEL VARIATIONS

XL53C (883cc)
XLH-883/XLH-883 HUGGER/
XLH-883C
XLH-1200/XLH-1200C/XLH-1200S

MODEL FLH

- **ENGINE** OHV V-TWIN
- **CAPACITY** 88CU. IN.

MODEL VARIATIONS

FLHR/FLHRCI

MODEL FLHT

- **ENGINE** OHV V-TWIN
- **CAPACITY** 88CU. IN.

MODEL VARIATIONS

FLHT/FLHTCI/FLHTCUI

MODEL FXD

- **ENGINE** OHV V-TWIN
- **CAPACITY** 88CU. IN.

MODEL VARIATIONS

FXD/FXDS/FXDWG/FXDL

1999 FXDS

SOFTAILS

- **ENGINE** OHV V-TWIN
- **CAPACITY** 80CU. IN.

MODEL VARIATIONS

FXST/FXSTS/FXSTC/FLSTS/
FLSTF/FLSTC

1999 FLSTF FAT BOY

- **COLORS**
WIDE RANGE OF STANDARD COLORS.

2000

Revised Softail range, with Twin Cam engine introduced across the range in a revised chassis. FXSTD is new Softail model.

MODEL XL/XLH

- **ENGINE** OHV V-TWINS
- **CAPACITY** 883cc & 1200cc

MODEL VARIATIONS

XL53C (883cc)
XLH-883/XLH-883 HUGGER/
XLH-883C
XLH-1200/XLH-1200C/XLH-1200S

MODEL FLH

- **ENGINE** OHV V-TWIN
- **CAPACITY** 88CU. IN.

MODEL VARIATIONS

FLHR/FLHRCI

MODEL FLHT

- **ENGINE** OHV V-TWIN
- **CAPACITY** 88CU. IN.

MODEL VARIATIONS

FLHT/FLHTCUI/FLTRI/

MODEL FXD

- **ENGINE** OHV V-TWIN
- **CAPACITY** 88 CU. IN.

MODEL VARIATIONS

FXD/FXDX/FXDWG/FXDL

SOFTAILS

- **ENGINE** OHV V-TWIN
- **CAPACITY** 88 CU. IN.

MODEL VARIATIONS

FXST/FXSTB/FXSTD/FXSTS/
FLSTC/FLSTF

- **NEW FEATURES** FOUR-PISTON CALIPER BRAKES ON ALL MODELS
FULLY ADJUSTABLE FRONT AND REAR SUSPENSION ON FXDX

- **COLORS**
WIDE RANGE OF STANDARD COLORS.

2001

Introduction of Super Glide T-Sport variant, effectively a Super Glide wih handlebar fairing and panniers.

MODEL XL/XLH

- ENGINE OHV V-TWIN
- CAPACITY 883cc & 1200cc

MODEL VARIATIONS

XL53C (883cc)
XLH-883/XLH-883 HUGGER/
XLH-883C
XLH-1200/XLH-1200C/XLH-1200S

2001 XLH-883

MODEL FLH

- ENGINE OHV V-TWIN
- CAPACITY 88CU. IN.

MODEL VARIATIONS

FLHR/FLHRCI

MODEL FLHT

- ENGINE OHV V-TWIN
- CAPACITY 88CU. IN.

MODEL VARIATIONS

FLHT/FLHTCUI/FLTRI/

MODEL FXD

- ENGINE OHV V-TWIN
- CAPACITY 88CU. IN.

MODEL VARIATIONS

FXD/FXDX/FXDWG/FXDL

2001 FXDX

SOFTAILS

- ENGINE OHV V-TWIN
- CAPACITY 88CU. IN.

MODEL VARIATIONS

FXST/FXSTB/FXSTD/FXSTS/
FLSTC/FLSTF

- NEW FEATURES FOUR-PISTON CALLIPER BRAKES ON ALL MODELS FULLY ADJUSTABLE FRONT AND REAR SUSPENSION ON FXDX

- COLORS
WIDE RANGE OF STANDARD COLORS.

2002

Introduction of flat-track inspired 883R. Introduction of all-new 1130cc VRSCA V-Rod, one of the most radical machines ever built by Harley-Davidson. Features a completely new, liquid-cooled, 60° V-twin engine in a cruiser-style chassis. Smaller "bullet" indicators used across the range.

MODEL XL/XLH

- ENGINE OHV V-TWIN
- CAPACITY 883cc & 1200cc

MODEL VARIATIONS

XL53C (883cc)
XLH-883/XLH-883 HUGGER/
XL-883R
XL-1200C/XL-1200S

MODEL FLH

- ENGINE OHV V-TWIN
- CAPACITY 88CU. IN.

MODEL VARIATIONS

FLHT/FLHTCUI/FLTRI

FLHTCUI

MODEL FXD (DYNAS)

- ENGINE OHV V-TWIN
- CAPACITY 88CU. IN.

MODEL VARIATIONS

FXD/FXDX/FXDWG/FXDL

2002 FXDL

SOFTAILS

- ENGINE OHV V-TWIN
- CAPACITY 88CU. IN.

MODEL VARIATIONS

FXST/FXSTB/FXSTE/FXSTD/
FLSTF/FLSTC/FLSTS

2002 FSXTB

2003

Centennial color schemes with special paint and anniversary badges across the range.

MODEL XL/XLH

- ENGINE OHV V-TWIN
- CAPACITY 883cc & 1200cc

MODEL VARIATIONS

XL53C (883cc)
XLH-883/XLH-883 HUGGER/ XL-883R
XL-1200C/XL-1200S

2003 XLH883R

MODEL FLH

- ENGINE OHV V-TWIN
- CAPACITY 88CU. IN.

MODEL VARIATIONS

FLHR/FLHRCI

MODEL FLHT

- ENGINE OHV V-TWIN
- CAPACITY 88CU. IN.

MODEL VARIATIONS

FLHTCUI

MODEL FXD (DYNAS)

- ENGINE OHV V-TWIN
- CAPACITY 88CU. IN.

MODEL VARIATIONS

FXD/FXDX/FXDWG/FXDL/
FXDXT

SOFTAILS

- ENGINE OHV V-TWIN
- CAPACITY 88CU. IN.

MODEL VARIATIONS

FXST/FXSTB/FXSTDI/FXSTS/
FLSTFI/FLSTCI/FLSTS

V-ROD

- ENGINE DOHC V-TWIN
- CAPACITY 1130cc

MODEL VARIATIONS

VRSCA

2003 VRSCA V-ROD

2004

New chassis and uprated engines for the Sportster range. CVO models use bigger 95cu. in. (1550cc) and 103cu. in. (1690cc) engines.

SPORTSTERS (XL)

- ENGINE OHV V-TWIN
- CAPACITY 883cc & 1200cc

MODEL VARIATIONS

XL883/XL883C/XL1200C/
XL1200R

TOURERS (FL)

- ENGINE OHV V-TWIN
- CAPACITY 88CU. IN. (1450cc)

MODEL VARIATIONS

FLHR/FLHRC/FLHRS/FLHT/
FLHTC/FLHTC/FLHTCU

DYNAS (FXD)

- ENGINE OHV V-TWIN
- CAPACITY 88CU. IN. (1450cc)

MODEL VARIATIONS

FXD/FXDL/FXDWG/FXDX

SOFTAILS (ST)

- ENGINE OHV V-TWIN
- CAPACITY 88CU. IN. (1450cc)

MODEL VARIATIONS

FLSTC/FLSTF/FXST/FXSTB/
FXSTD/FXSTS

V-ROD

- ENGINE DOHC V-TWIN
- CAPACITY 1130cc

MODEL VARIATIONS

VRSCDX/VRSCF

2005

New Softail models and CVO versions of V-Rod, Fat Boy, and Electra Glide.

SPORTSTERS (XL)

- ENGINE OHV V-TWIN
- CAPACITY 883cc & 1200cc

MODEL VARIATIONS

XL883/XL883C/883L/883R
XL1200C /XL1200R

TOURERS (FL)

- ENGINE OHV V-TWIN
- CAPACITY 88CU. IN. (1450cc)

MODEL VARIATIONS

FLHR/FLHRC/FLHRS/FLHT/
FLHTC/ FLHTCU/FLTR

DYNAS (FXD)

- ENGINE OHV V-TWIN
- CAPACITY 88cu. in. (1450cc)

MODEL VARIATIONS

FXD/FXDL/FXDW G/FXDX/FXDC

SOFTAILS (ST)

- ENGINE OHV V-TWIN
- CAPACITY 88cu. in. (1450cc)

MODEL VARIATIONS

FLSTC/FLSTSC/FLSTN/FLSTF/
FXST/FXSTB/FXSTD/FXSTS

V-ROD

- ENGINE DOHC V-TWIN
- CAPACITY 1130cc

MODEL VARIATIONS

VRSCB/VRSCA

2006

Dyna range re-engineered with new frame and a six-speed gearbox, includes a 35th anniversary Super Glide.

SPORTSTERS (XL)

- ENGINE OHV V-TWIN
- CAPACITY 883cc & 1200cc

MODEL VARIATIONS

XL883/XL883L/XL883C/XL883R/
XL1200C/XL1200R/XL1200L

TOURERS (FL)

- ENGINE OHV V-TWIN
- CAPACITY 88cu. in. (1450cc)

MODEL VARIATIONS

FLHR/FLHRS/FLHRC/FLTR/
FLHX/FLHT/FLHTC/FLHTCU

DYNAS (FXD)

- ENGINE OHV V-TWIN
- CAPACITY 88cu. in. (1450cc)

MODEL VARIATIONS

FXD/FXDC/FXD35/FXDB/FXDL/
FXDWG

SOFTAILS (ST)

- ENGINE OHV V-TWIN
- CAPACITY 88cu. in. (1450cc)

MODEL VARIATIONS

FXST/FXSTB/FXSTS/FXSTS/
FXSTD/FLST/FLSTF/FLSTN/
FLSTSC/FLSTC

V-ROD

- ENGINE DOHC V-TWIN
- CAPACITY 1130cc

MODEL VARIATIONS

VRSCA/VRSCD/VRSCR

2007

New 96-cu. in. (1584-cc) engine for Dyna, Softail and Touring models, now all with 6-speed gearbox. V-Rod capacity increases to 1250cc. 50th anniversary Sportster.

SPORTSTERS (XL)

- ENGINE OHV V-TWIN
- CAPACITY 883cc & 1200cc

MODEL VARIATIONS

XL883/XL883L/XL883R/XL883C/
XL50/XL1200L/XL1200R/XL1200C

TOURERS (FL)

- ENGINE OHV V-TWIN
- CAPACITY 96cu. in. (1584cc)

MODEL VARIATIONS

FLHX/FLHR/FLHRS/FLHRC/
FLHT/FLHTC/FLHTCU/FLHR

DYNAS (FXD)

- ENGINE OHV V-TWIN
- CAPACITY 96cu. in. (1584cc)

MODEL VARIATIONS

FXD/FXDB/FXDL/FXDC/FXDWG

SOFTAILS (ST)

- ENGINE OHV V-TWIN
- CAPACITY 96cu. in. (1584cc)

MODEL VARIATIONS

FXSTC/FLSTN/FLSTSC/FLSTC/
FXST/FXSTB/FXSTD

V-ROD

- ENGINE DOHC V-TWIN
- CAPACITY 1250cc

MODEL VARIATIONS

VRSCAW/VRSCX/VRSCDX/
VRSCD/VRSCDX

2008

105th anniversary models, include the new, custom-look Rocker and Rocker C Softail models. And Europe gets the XR1200.

SPORTSTERS (XL)

- ENGINE OHV V-TWIN
- CAPACITY 883cc & 1200cc

MODEL VARIATIONS

XL883/XL883L/XL883C/XL1200L/
XL1200R/XL1200C

TOURERS (FL)

- ENGINE OHV V-TWIN
- CAPACITY 96cu. in. (1584cc)

MODEL VARIATIONS

FLHR/FLHRC/FLHX/FLHTR/
FLHT/FLHTC/FLHTCU

DYNAS (FXD)

- ENGINE OHV V-TWIN
- CAPACITY 96cu. in. (1584cc)

MODEL VARIATIONS

FXD/FXDB/FXDL/FXDC/FXDWG/
FXDL/FXDWC

SOFTAILS (ST)

- ENGINE OHV V-TWIN
- CAPACITY 96cu. in. (1584cc)

MODEL VARIATIONS

FXSTB/FXSRC/FLSTF/FLSTN/
FLSTC/FLSTSB/FXCW/FXCWC

V-ROD

- ENGINE DOHC V-TWIN
- CAPACITY 1250cc

MODEL VARIATIONS

VRSCAW/VRSCD/VRSCDX

2009

Tourers get a new frame, and there is also a trike.

SPORTSTERS (XL)

- ENGINE OHV V-TWIN
- CAPACITY 883cc & 1200cc

MODEL VARIATIONS

XL883L/XL883N/XL883C/
XL1200L/XL1200N/XL1200C/
XR1200

TOURERS (FL)

- ENGINE OHV V-TWIN
- CAPACITY 96cu. in. (1584cc)

MODEL VARIATIONS

FLHR/FLHRC/FLHT/FLHTC/
FLHTCU/FLHX/FLTR

DYNAS (FXD)

- ENGINE OHV V-TWIN
- CAPACITY 96cu. in. (1584cc)

MODEL VARIATIONS

FXD/FXDB/FXDC/FXDF/FXDL

SOFTAILS (ST)

- ENGINE OHV V-TWIN
- CAPACITY 96cu. in. (1584cc)

MODEL VARIATIONS

FXSTC/FXSTB/FXCWC/
FXCWFLSTB/FLSTN/FLSTF/
FLSTC

V-ROD

- ENGINE DOHC V-TWIN
- CAPACITY 1250cc

MODEL VARIATIONS

VRSCAW/ VRSCDX/VRSCF

TRI-GLIDE

- ENGINE OHV V-TWIN
- CAPACITY 103cu. in.

MODEL VARIATIONS

FLHTCUTG

MODEL CVO

- ENGINE OHV V-TWIN
- CAPACITY 110cu. in.

MODEL VARIATIONS

FXSTS/FXDF/FLTR/FLHTCU

2010

Detail changes across the range. Police, Firefighter, and Shrine special editions available.

SPORTSTERS (XL)

- ENGINE OHV V-TWIN
- CAPACITY 883cc & 1200cc

MODEL VARIATIONS

XL883L/XL883N/FORTY-EIGHT/
XL1200L/XL1200N/XL1200C/
XR1200

2010 XR1200X

TOURERS (FL)

- ENGINE OHV V-TWIN
- CAPACITY 96cu. in.(1584cc)

MODEL VARIATIONS

FLHR/FLHRC/FLHTC/FLHTK/
FLHTCU/FLHX/FLTRX

DYNAS (FXD)

- ENGINE OHV V-TWIN
- CAPACITY 96cu. in.(1584cc)

MODEL VARIATIONS

FXD/FXDB/FXDC/FXDF/FXDWG

SOFTAILS (ST)

- **Engine** OHV V-twin
- **Capacity** 96cu. in. (1584cc)

MODEL VARIATIONS

FXSTC/FXCWC/FLSTB/FLSTN/
FLSTSB/FLSTN/FLSTFB/FLSTF/
FLSTC

V-ROD

- **Engine** DOHC V-twin
- **Capacity** 1250cc

MODEL VARIATIONS

VRSCAW/ VRSCDX/VRSCF

2010 VRSCDX

TRI-GLIDE

- **Engine** OHV V-twin
- **Capacity** 103cu. in.

MODEL VARIATIONS

FLHTCUTG/FLHXXX

MODEL CVO

- **Engine** OHV V-twin
- **Capacity** 110cu. in.

MODEL VARIATIONS

FLST/FLHX/FLHTCU/FXDF

2011

The 2011 range included the new "Super Low" Sportster 883 and the Road Glide Ultra featuring the 130cu. in. engine.

SPORTSTERS (XL)

- **Engine** OHV V-twin
- **Capacity** 883cc & 1200cc

MODEL VARIATIONS

XL883L/XL883N/Forty-Eight/
XL1200L/XL1200N/XL1200C/
XR1200X

TOURERS (FL)

- **Engine** OHV V-twin
- **Capacity** 96cu. in. (1584cc)

MODEL VARIATIONS

FLHR/FLHRC/FLHTC/FLHTCU/
FLHTK/FLHX/FLTRU/FLTRX

DYNAS (FXD)

- **Engine** OHV V-twin
- **Capacity** 96cu. in. (1584cc)

MODEL VARIATIONS

FXDB/FXDC/FXDF/FXDWG

SOFTAILS (ST)

- **Engine** OHV V-twin
- **Capacity** 96cu. in. (1584cc)

MODEL VARIATIONS

FXS/FXCWC/FLSTFB/FLSTF/
FLSTC/FLSTN/FLSTSB

V-ROD

- **Engine** DOHC V-twin
- **Capacity** 1250cc

MODEL VARIATIONS

VRSCDX/VRSCF

TRI-GLIDE

- **Engine** OHV V-twin
- **Capacity** 103cu. in.

MODEL VARIATIONS

FLHTCUTG/FLHXXX

MODEL CVO

- **Engine** OHV V-twin
- **Capacity** 110cu. in.

MODEL VARIATIONS

FLTRU/FLHTCU/FLHX/FLST

2012

New 103cu. in. (1690cc) engine on Softail and Touring models, and most Dynas. New Seventy Two Sportster.

SPORTSTERS (XL)

- **Engine** OHV V-twin
- **Capacity** 883cc & 1200cc

MODEL VARIATIONS

XL883L/XL883N/XL1200V
Seventy Two/XL1200X Forty-
Eight/XL1200C/XL1200N/
XR1200X

TOURERS (FL)

- **Engine** OHV V-twin
- **Capacity** 96cu. in. (1584cc)

MODEL VARIATIONS

FLHR/FLHRC/FLHTC/FLHTCU/
FLHTK/FLHX/FLTRX/FLHTRU

DYNAS (FXD)

- **Engine** OHV V-twin
- **Capacity** 96cu. in. (1584cc)

MODEL VARIATIONS

FXD/FXDB/FXDL/FXDC/FXDWG

SOFTAILS (ST)

- **Engine** OHV V-twin
- **Capacity** 96cu. in. (1584cc)

MODEL VARIATIONS

FLS/FLSTN/FLSTFB/FLSTF/
FLSTC/FLSTN/FXS

V-ROD

- **Engine** DOHC V-twin
- **Capacity** 1250cc

MODEL VARIATIONS

VRSCDX/VRSCF

TRIKE

- **Engine** OHV V-twin
- **Capacity** 103cu. in.

MODEL VARIATIONS

FLHTCUTG

MODEL CVO

- **Engine** OHV V-twin
- **Capacity** 110cu. in.

MODEL VARIATIONS

FLTRX/FLHX/FLST/FLHT

2013

The company celebrates their 110th anniversary with special detailing and graphics. The only significant new model is the FXBSE Breakout, a custom-style Softail with CVO finishes and an 1802cc engine.

SPORTSTERS (XL)

- **Engine** OHV V-twin
- **Capacity** 883cc & 1200cc

MODEL VARIATIONS

XL883L/XL883N/XL1200C/
XL1200X/XR1200V

TOURERS (FL)

- **Engine** OHV V-twin
- **Capacity** 103cu. in. (1690cc) & 110cu. in. (1802cc)

MODEL VARIATIONS

FLHR/FLHRC/FLHX/FLTRX/
FLTRU/FLHTC/FLHTCU/FLHTK

DYNAS (FXD)

- **Engine** OHV V-twin
- **Capacity** 96cu. in. (1584cc) & 103cu. in. (1690cc)

MODEL VARIATIONS

FXDB/FXDC/FXDWG/FXDF/FLD

SOFTAILS (ST)

- **Engine** OHV V-twin
- **Capacity** 103cu. in. (1690cc) & 110cu. in. (1802cc)

MODEL VARIATIONS

FLSTF/FLSTFB/FXS/FLSTN/
FLSTC/FLS/FXSBSE

V-ROD

- **Engine** DOHC V-twin
- **Capacity** 76.25cu. in. (1250cc)

MODEL VARIATIONS

VRSCDX/VRSCF

TRIKE

- **Engine** OHV V-twin
- **Capacity** 103cu. in. (1690cc)

MODEL VARIATIONS

FLHTCUTG

Harley-Davidson Model Designations

Harley-Davidson introduced its lettering system in 1909 to differentiate between models. No numbers are included, but up until 1916 the prefix number was four fewer than the actual year (so 1912 models were prefixed by the number 8) and from 1916 the year was used as the prefix (so 1922 models were prefixed by 22). Some letters are placed first in the designation to describe engine type, some are suffixes that add extra information about the engine, and others are purely descriptive. The system doesn't really start to make sense until the 1920s, so no designations from before this date are included in this rough guide.

KEY TO ENGINE TYPE:
sv = side-valve ohv = overhead-valve
ts = two-stroke ioe = inlet-over-exhaust

MODEL A PEASHOOTER

A: DESIGNATIONS

A: (1926–30) sv 21cu. in. "Peashooter" single with magneto

A: (1960–65) ts 165cc A Topper scooter

A: Army version (i.e. 1942 WLA)

A: Ohv version of the A and B singles (i.e. 1926 AA and BA)

B: DESIGNATIONS

B: (1926–28) sv 21cu. in. "Peashooter" single with battery ignition and lights

B: (1955–59) 125cc ts single "Hummer"

BT: (1960–66) 165cc ts single

C: DESIGNATIONS

C: (1929–34) sv/ohv 30.5cu. in. single

C: Custom/classic/cafe

CH: Magneto ignition Sportster

C: Canadian army version (i.e. 1942 WLC)

D: DESIGNATIONS

D: (1929–31) sv 45cu. in. twin

D 74cu. in. version of J series (i.e. 1921 JD)

D: "Dyna" frame (i.e. FXDWG)

D: High-compression version (i.e. 1932 RLD)

DG: Disc Glide (i.e. FXDG)

E: DESIGNATIONS

E: (1936–52) 61cu. in. ohv twin "Knucklehead" or "Panhead"

E: Electric starting (i.e. 1974 FXE)

E: Police/Traffic Combination engine (i.e. 1954 FLE)

F: DESIGNATIONS

F: (1914–25) ioe 61cu. in. magneto ignition twin

F: (1941–78) ohv 74cu. in. twin (note: since 1978 the F initial has also been used on 80cu. in. and 88cu. in. engines)

F: Battery ignition flat twin (i.e. 1922 WF)

F: Footshift

G: DESIGNATIONS

G: (1933–73) Servi-Car

H: DESIGNATIONS

H: Larger engine version of existing model (i.e. 1936 VH, 1955 KH)

H: More powerful engine (i.e. 1959 XLH)

I: DESIGNATIONS

I: Fuel injection (i.e. 1995 FLHTCUI)

J: DESIGNATIONS

J: (1915–29) 61cu. in. ioe twin

J: Magneto ignition flat-twin (i.e. 1922 WJ)

K: DESIGNATIONS

K: More powerful KH model (i.e. 1956 KHK)

K: (1952_53) 45cu. in. sv twin

L: DESIGNATIONS

L: Higher compression engine (i.e. 1936 EL)

M: DESIGNATIONS

M: (1965–72) 50cc/65cc ts single

N: DESIGNATIONS

N: Nostalgia (i.e. 1993 FLSTN)

P: DESIGNATIONS

P: Police model

R: DESIGNATIONS

R: (1932–36) 45cu. in. sv twin

R: Rubber-mounted engine on FX model (i.e. 1983 FXR)

R: Racing model (i.e. 1957 KR)

R: Race-derived model (i.e. 1984 XR1000)

S: DESIGNATIONS

S: (1926–30) 21cu. in. ohv racer

S: (1948–52) 125cc ts single

ST: (1953–59) 165cc ts single

S: Sport (i.e. FXS)

S: Sidecar model (i.e. 1926 JDS)

S: Softail

T: DESIGNATIONS

T: (1921) twin-cylinder racer

T: Touring version (i.e. 1978 XLT)

TT: Road-race version of competition bike (i.e. XRTT)

U: DESIGNATIONS

U: (1937–48) sv 74cu. in. twin

U: Restricted version (i.e. STU and AU Topper)

U: Ultra accessory package

V: DESIGNATIONS

V: (1930–36) sv 74cu. in. twin

V: (1994–) dual overhead cam 61cu. in. twin for Superbike racing (i.e. 1994 VR1000)

W: DESIGNATIONS

W: (1919–23) sv 36cu. in. flat-twin

W: (1937-51) sv 45cu. in. twin

WG: Wide-Glide (i.e. 1980 FXWG)

X: DESIGNATIONS

XL: (1957–) 55/61/74cu. in. ohv twin Sportster

XA: (1944) 45cu. in. sv flat-twin for the US army

X: Super Glide (i.e. 1971 FX)

Z: DESIGNATIONS

Z: (1973) 90cc ts single

XR1000 SPORTSTER

Index

Note: Models are listed both under their designation and under their name. So, for example, the AH Topper comes under "A" as well as under "T." No models from the catalog are included except for those included in the text that summarizes each year.

1915 KT BOARD RACER

E

KNUCKLEHEAD ENGINE

F

G

H

I

1941 WLD

1941 WLD

1941 W<small>LD</small>